How to Become an Ecological Consultant

*How to Get the Skills, Knowledge and Experience
You Need for a Job in Ecological Consultancy
in the UK*

By

Susan M Searle BSc (Hons), PGDip (Ecology), MIEEM
Senior Ecologist and Managing Director
Acorn Ecology Ltd
www.acornecology.co.uk

This book is dedicated to two inspirational people:

Michael Woods who died aged 61 in January 2010. He was the chairman of The Mammal Society for 6 years and on the Council for 27 years. He was a travel writer and instrumental in the setting up of the National Cycle Network. As one of the UK's top experts in badgers and dormice he helped develop dormouse tubes as a survey technique and, just before his death, wrote a book on badgers for The Mammal Society. He carried out voluntary work until the end, mainly with dormice. He ran his own consultancy and was always supportive and helpful to me in my career, especially when I was in need of expert advice. He was also a very nice person and a good friend.

Warren Cresswell who died aged 47 in December 2009. He taught me many of the survey skills I needed for my new career during my MSc course and inspired me to start up my own consultancy. He provided willing support regularly during the early years of my career. He was one of the top ecologists in the UK and an extremely good person.

They are sadly missed, both by me,
and the rest of the ecology world.

Sue Searle

THANK YOU
To Dr Nigel Massen for finally inspiring me
to get this book written and published and
helping me prepare the manuscript.

About the Author

Sue Searle started her career in nursing and midwifery but later in life decided to follow her passion for wildlife. At 39 she started her degree in Biological Sciences at University of Exeter and completed it in 2001. She then studied a Postgraduate Diploma in Ecology, Conservation and Habitat Management at University of Bristol.

In 2003 Sue set up Acorn Ecology Ltd with no experience of ecological consultancy apart from her studies. Through the help and support of others and by 'just doing it' the ecological consultancy business is now thriving and provides, amongst other things, ecological training for people wanting to gain skills and knowledge to get into ecological consultancy.

Sue's philosophy is to help as many people as possible to fulfill their career aspirations and to help produce quality, well-trained ecologists in order to ensure that wildlife can thrive despite the pressures it experiences from human beings.

Sue also runs courses and writes on personal development (www.thelifeyouchoose.net) and is an accomplished artist (www.artysueart.co.uk). She runs Pioneer Ecology with Simon Trevenna which takes people abroad to learn about ecology (www.pioneerecology.com).

This book is not intended to be a definitive guide that will work for all but as a source of inspiration to make a start on gaining the skills and knowledge to gain employment in ecological consultancy. Each job advertised will require a different set of skills and each person will have their own interests and passions which they can bring to a job. The information and suggestions in this book are those of the author and it is up to you whether you consider the advice to be suitable for your own personal situation.

Table of contents

Introduction   1

Chapter 1 Three career paths in ecology   5

Chapter 2 The professionalization of ecology   23

Chapter 3 What is an ecological consultant?   36

Chapter 4 Applying knowledge, expanding skills   48

Chapter 5 Volunteering   64

Chapter 6 Demonstrating your skills   79

Chapter 7 Your Curriculum Vitae (CV) and
           covering letter   92

Chapter 8 Successful interview technique   107

Chapter 9 First job and beyond   119

Chapter 10 Develop a specialism   136

Chapter 11 Final thoughts   144

Appendix 1 Common Laws and useful websites   147

Appendix 2 Example of CV layout   155

Appendix 3 Life Planner blank   156

Bibliography   157

Introduction

My early career

Pounding the wards of a busy teaching hospital in Bristol in my starched apron and hat, I would never have imagined that 30 years later I would be a professional ecologist running my own successful consultancy.

After training for 4 years to become a nurse and a midwife the daily routine of tending the sick, injured and dying in the clinical environment, or delivering babies, was a world apart from my working day now. I was bestowed great responsibility at a very young age as a trainee nurse - in charge of my first ward on night duty at the age of 18 seems pretty unbelievable now. Some of the stories I could tell about that phase of my working life would horrify you, and without having experienced nursing you would never be able to imagine them, so I will spare you!

Gaining the trust of patients, from small children to young men of my age, from the frail elderly to those horribly injured; I had to learn quickly how to deal with people. If you are about to stick a needle into someone's arm, or deliver their first baby, they need to totally trust you and feel confident in your abilities. Dealing with colleagues was also sometimes a challenge as we were often extremely tired or working in stressful situations. From porters to senior consultants, nursing assistants to senior nurses, all had to be dealt with in different ways to ensure we worked together as a team. I thank my nursing experience for such valuable skills – taking responsibility, people skills, working long hours, dealing with stressful and difficult situations and working with confidence. All have held me in good stead for the rest of my working life.

A typical week now

There is no typical day, or even season, in my career now. Ecological consultancy has got to be one of the most interesting, varied, intellectual and challenging jobs around. Last week, for instance, I closed a badger sett, carried out a dawn and two dusk bat surveys, did a ditch dipping session with some local children, found a dormouse in a hedge and explained to a client the implications of having a bat maternity roost in their loft if they wanted to demolish their house. I also wrote several reports and, as part of the running of the business, spoke to clients, did quotes, accounts, correspondence, marketing, paid bills, VAT and wages and helped my staff with their work. Quite a week! Each one is just as interesting, challenging and varied. I would not give up this career for the world, it's great, and it certainly doesn't seem like work most of the time. I have been an ecological consultant since 2003 when I set up my own business, Acorn Ecology Limited.

Why ecological consultancy?

One burning question you may have at this point is how or why did I decide to become an ecologist? I will cover this later as we explore what might motivate you to become an ecologist. Simply, I was always interested in wildlife, particularly plants, mammals, reptiles and insects and I love being outdoors. My early years were spent in Africa and as a child my mother could not get me to come indoors!

When I left school in 1976 ecology was not on the radar as a career. It was not until much later, in the 1990s, that I realised I could make a career of wildlife and what I needed to do to get there. Ecological consultancy became a more precise goal towards the end of my first degree, but more on that later.

What will you gain from this book?

As a person who is presumably thinking of becoming an ecological consultant I hope that by reading this book you will gain some insight into what the job entails and what challenges you might expect.

This book is written to help you make a start in a career as an ecological consultant. You may have just completed a degree or may be looking for a change in career like I did. Finding a job in this sector can seem hard, especially if you have no experience. But if you put your skills and knowledge in place bit by bit you will succeed.

This book will tell you what you need to know to get started as a professional ecologist; what academic skills you will need; how to learn the valuable skills you'll need to work with wildlife; how to gain the right kind of practical experience and how to demonstrate your knowledge in the right way to impress employers. I will also cover how to present your CV and yourself, when job hunting.

Chapter 1

∞∞∞∞

Three career paths in ecology

It is at this point that I would like to talk about the three different paths that you can take to build a career in ecology. It took me quite a while to realise that there are actually very different paths to take as an ecologist, and that the difference between them is not generally discussed much between ecologists when they're starting out. I want to point it out now to make sure you decide which way you want to go before you go too far in one direction and find you have to back-pedal. All three are very enjoyable, but they differ in the type of challenge they offer, the level of intellectual involvement, maintenance of professional standards and, last but not least, they can differ vastly in salary potential.

This book is focused on career path No. 3 - ecological consultancy and how to get into it. I shall explain what it is and the difference between that and the ecology career paths outlined below. I hope that you will then be able to at least decide which path you want to take your career and make a start in the right direction.

These career paths are not mutually exclusive but you will need to plan your education, skills and experiences around the one you wish to pursue a career in.

Career path No. 1 – conservation

I am not sure how to describe one of the paths apart from calling it 'conservation'. There are a myriad of great careers in conservation - I have had a few myself. Conservation jobs are usually for charities and trusts, local authorities, government organisations, wildlife trusts, small conservation organisations, museums, zoos and campaigning groups. Jobs in this area can be found both here in the UK and abroad. These jobs suit certain types of people and can be highly varied and rewarding.

As many of these jobs are funded by grants or fundraising it is nearly always the case that these jobs are either very poorly paid, based on short-term contracts, or even voluntary. These jobs are mostly non-income generating. From speaking to many people working for conservation charities in particular, there can often be stressful times when funding runs short or is completely used up and your job has to go. If you get a fixed-term contract or project officer job then you can't expect it to be extended. My local wildlife trust went through a serious shrinkage a few years ago and dozens of jobs were axed due to lack of funds. Staff numbers seem to fluctuate over time according to the success of their fundraisers and the economy in general.

Conservation jobs

Many jobs available in the UK to do with wildlife and ecology seem to be 'conservation' jobs. I think these jobs are easier to get into if you lack experience but there is still stiff competition. Advising on the sorts of skills you might need for these conservation jobs is difficult as each one is so different. Organisational abilities, methodical working, teamwork, good people skills and a good level of literacy are the most common skills required but most will expect a keen interest in wildlife, experience and a relevant degree.

One huge area of conservation is site management. This includes wardens, countryside rangers, site managers, reserves officers and reserves assistants. Skills for this sort of job start with practical habitat management and a smattering of public awareness work, depending on the site.

Familiarity with the use of chain saws, loppers, handsaws, weed killer sprays and other hand tools are part of the training and certificates for their safe use can be gained. Big land-owning conservation charities such as the National Trust, RSPB and the wildlife trusts are the main employers.

Nature reserves need to be managed to keep the habitats the same or to enhance the biodiversity on site and so these jobs tend to be for more practical hands-on people who like to work outdoors. Activities like scrub-bashing, clearing bracken or reeds and mowing can often take up large parts of your working time. Reserves management often involves managing teams or work parties and as well as preparing management plans, monitoring wildlife and habitat change, talking to the public and holding events. In many cases the job may not actually entail doing the practical conservation work yourself but managing groups that do. Other conservation jobs such as lobbying, public awareness, managing teams of volunteers, working with schools or visitors to centres, may be more focused on your people skills, your ability to speak publically, your negotiation skills, maybe even your command of foreign languages. Each post will have its own unique role and set of skills that you will need to fulfill. Pretty interesting work but unfortunately not usually well paid.

Many people that work in the area of site management stay for many years and I think it is often the case that the job can be adapted to your skills to a certain extent – for example if you enjoy working with children you can do events for them specifically. A high degree of job satisfaction can be gained from seeing a site change for the better over time under your care and engaging with an enthusiastic public. During this time you can build up expertise in wildlife that you are dealing with, for example you could become a botanist, birder or reptile expert which would stand you in good stead if you then decided to change direction and go into ecological consultancy.

If this does sound like your cup of tea then BTCV and other organisations offering practical skills, or volunteering for the above named conservation charities on their reserves, is a great way to start.

For the other posts it is likely that you will need to be prepared to move on regularly, gradually building up your skills. With this pathway it is harder to see the direction you might be going in with your career as it will be heavily reliant on the posts that come available.

Career path No. 2 – academia

Academia is a well-established path you can take after your degree - there is a whole career structure waiting for you and you can pursue jobs literally anywhere in the world. Academics are the ones that do the research that helps everyone in ecology work effectively and make sure that the advice we give has a scientific foundation. They discover, through their research, what works and what doesn't in terms of habitat and species management, population structure and much more. They also study species ecology and a myriad other things including all the genetics studies that have shed so much light on how species interact and their evolutionary history. They are the ones that we listen to avidly when at conferences, when they describe their research findings - this helps us hone our knowledge and advice.

An academic career follows a pathway and takes many years to fulfill, and goes something like this:

After your first degree you are expected to do a Master's degree (although you can go straight into a PhD), a word of caution here, make sure you choose one that will actually serve you in the future. This will then be closely followed by a PhD, again, make sure the topic you spend three or four years of your life studying will serve you in the future. If you have less than a 2:1 in your degree then you will probably not be considered for a PhD in the first place. A PhD typically lasts for three/four years, during this time will probably be expected to finish in, and be funded for, three years. You can expect to earn in the region of £12,000 per year. If you overrun into a fourth year while writing-up your thesis, you may need to work to support yourself at the end.

After your PhD you will carry out post-doctoral research, usually a two-year contract to work on a specific task. You will be expected to produce several papers, present at conferences and join in with working groups on your specialist topic. You may do two or three such post-docs before moving onto a Fellowship or Readership post where you would typically spend 5–10 years. By now you would expect to have more responsibility, some admin roles, have to find funding for research and have your own PhD students. You would also have some teaching responsibility.

Next you would look for a full-time lectureship post with a permanent contract. This has a lot more teaching involved, tutoring your own students, supervising PhD students, finding funding and you would have more admin duties and less time for research, maybe only having a chance to get away in the summer holidays for fieldwork. You might, at this stage be involved in some management of your own courses. Salaries are typically £40–50,000. You could stay in this position indefinitely.

After some experience you would become senior lecturer, or Director of Teaching, where you would have responsibility for other staff, could be managing teaching for a whole year group of undergraduates, have more teaching duties, more admin duties, you will have to find funding for research whilst not actually having time to do any yourself. Pay by now is £50–60,000 and this position can be held indefinitely.

Finally you reach the elevated position of Professor. For this you typically have to be nominated. You are working at a top level but with similar duties to senior lecturer, the difference is you will get a better salary and more responsibility for decisions within your department. You are now a senior academic; working at the top level of both your institution and internationally in your field.

Due to the scarcity of lecturer, senior lecturer and Professorial jobs, there is fierce competition at every stage. Career success is heavily dependent on research success (measured by the impact factor of your published work) and how much funding you have brought into your department. A lot of your job, especially as you progress, is about seeking funding for your research group (i.e. the post-docs and PhDs who work in your lab). It's that aspect of the job that puts off many a potential academic.

Career path No. 3 - ecological consultancy

Ecological consultants are paid by their clients, usually developers, to deliver advice on wildlife and conservation issues which they might have an impact on, usually while they are applying for planning permission or during a development. Projects that ecological consultants would be involved in range from minor barn alterations to new motorways – both would have to take account of their likely impact on wildlife and takes steps to mitigate it.

Ecological consultants are heavily involved in major strategic projects such as the London Olympics, Severn barrage, wind farms, high speed rail links and the channel tunnel railway. The ecological surveys and mitigation projects can sometimes take many years to complete or resolve. Ecologists are also often involved in habitat creation and re-creation, ongoing site management and monitoring, mitigating for habitat loss and assessing environmental impacts on the ecology of a site.

What we advise can have serious impacts on the development. For instance we may advise that works cannot be carried out in the summer; a hedgerow must be retained; or a barn cannot be converted due to being regionally important for a rare bat species. It is here that your paying client may not always be happy about the advice you give them! Luckily we have the law on our side for protecting wildlife and most clients are keen to find a way for their development and the natural world to coexist. We have considerable power to influence developers to do the right thing for wildlife.

Ecological consultancy is very much a profession now and I will explain this more in the next chapter. The emphasis of our role is therefore very different from conservation. It is very unlikely that your employer would sue you, for mismanaging your work on a nature reserve. Neither is it likely you would get into serious trouble if you under-performed in your role as a project officer. In conservation you may be accountable to your funders for reaching targets but the likelihood of you having to deal with legal issues or to be accountable for the advice you give is fairly remote.

How consultancy differs from conservation

Conservation jobs can involve some similar tasks to ecological consultancy but these are mostly related to surveying rather than providing professional advice. It is interesting to note here that many wildlife trusts now have ecological consultancies attached to them or employ their own in-house consultants and they are proving to be excellent income generators. However, the consultants that work in them, especially at senior level, do not, unfortunately, seem to command the higher salaries that purely commercial consultants can.

Many jobs in conservation involve carrying out some aspect of wildlife management; or maybe disseminating information to the public about a specific issue; they do not typically involve the level of decision-making, or professional standards and legal responsibility that an ecological consultant has to their clients.

Giving professional advice

Giving professional advice requires a fair bit of experience; not only making an assessment of impacts but also planning appropriate mitigation for any loss and dealing with the legal implications. Our advice needs to stand up or we can easily find ourselves with a bad reputation and clients that will never use us again. Bad advice could also lose us credibility with statutory wildlife bodies and local authority planning departments. Experience gives you more confidence, especially in what professional standards are expected in terms of survey effort, assessment of impacts and mitigation.

An in depth knowledge of the legislation relating to wildlife is essential to give proper advice. Similarly a good knowledge of specific species requirements for survey and mitigation is essential. It is unlikely that you will get the green light for your client if you have suggested some survey technique or mitigation that has never been proved to work. This means that you need to keep up to date with new findings and changes in legislation and for this being a member of the Institute of Ecology & Environmental Management (IEEM) and other professional bodies is very useful. Attending conferences, reading newsletters, going on training courses held by experts in their field and keeping up with the news generally is essential to keep ahead of the game. Continual Professional Development (CPD) is compulsory for IEEM members and is set at 20 hours a year – I regularly complete well over 100 hours of structured CPD (many hours of which are from attending conferences).

This might all sound a bit scary but actually it is all part of what makes the career interesting, challenging and intellectual. As we are working for paying customers and some survey work can be very extensive (and expensive), our level of earnings have the potential to be pretty impressive. I know several consultancy owners who are now millionaires (incidentally, they still work full time in consultancy as they enjoy it so much). I told one of my trainees that she could be a senior ecologist in 4 years, and she had seen the salaries in the adverts – but she did not believe me of course. However, with direction, focus, motivation and perseverance you can propel your career pretty quickly once you get that first job. We will be covering how you can get on the ladder in later chapters.

What does ecological consultancy involve?

I regard us more as applied scientists with part of our job being directly compiling data and interpreting it. In order to give accurate advice we need to have robust evidence that we have completed enough appropriate survey work to be sure of the species we are dealing with and what status the site is for each species (e.g. is it an important breeding site?). Much of our working life is therefore spent doing wildlife surveys for clients, especially in the summer. This seasonality often has negative impacts on workload, and trainees in particular may find it difficult to get anything but seasonal jobs at first.

Once we know which species we are dealing with and what is proposed for the site we then advise on the impacts of the development in the context of the legal protection of wildlife and habitats and ecosystem integrity. In some cases this initial survey and assessment can take months or even years as in the case of major projects. The channel tunnel rail link, for instance, which affected numerous wildlife sites, ancient woodlands, wetlands and other important features such as archaeological sites literally kept ecologists in work for years.

A major part of our job is the production of a report that summarises our survey methods, results and conclusions. This report provides specific mitigation advice, where needed, and legal advice, meant both for the client and the planner who is trying to decide whether the project is going to get planning permission.

Public consultation is part of the planning process, so our reports are in the public domain and do come under intense scrutiny if there is a dispute. They are often found on the online planning portals of the local authorities whilst the planning application is being decided. It is during this time that the statutory nature conservation organisations - SNCO (e.g. Natural England, Countryside Council for Wales and Scottish Natural Heritage, Environment and Heritage Service (Northern Ireland)) see the applications and can object on the grounds of lack of survey effort. There are survey protocols for most species now and if they are not followed rigidly then the planning application may be stalled. We often find that clients are reluctant to commission us to complete our survey work until the SNCO or the planners throw out the application due to lack of survey information (as we originally advised them).

Also at this stage you can get the public phoning you up to try to get you to give out more information about a development – often if something is being built near where they live. Some locals may, for instance, tell the local authority there is a badger sett on site or a bat roost so we may be asked to do a survey to check whether this is the case or not. The public can often mistake rabbit burrows for badger sett entrances and most people think that if bats fly around an area they must be roosting there too. Our professional advice can confirm or discount these types of claims to ensure that wildlife is protected, but also that developments can go ahead if all legal protection for wildlife is taken into account.

Variety of work

We don't just do work for developers although this does seem to be the main work for most consultancies and some work solely in this field. I have worked on many projects where I have just provided wildlife information (e.g. factsheets, information leaflets, wildlife surveys with enhancement ideas). I have also worked on strategic projects such as a review of our county Biodiversity Action Plan. Jobs have ranged from research projects, writing informative web sites, holding public events, running training courses, talking to or taking school children out, photographic monitoring, preparing biodiversity actions plans for quarries and parishes, preparing wildlife policies for large landowners, giving advice to landowners on wildlife enhancement, schools activity packs and interpretation board design.

Other random projects, such as looking after sick and injured bats, offer a diversion from the day-to-day routine of survey, report, survey, report. Many of these non-survey activities occupy our winter months when most survey work is not possible. The scope of the work of each consultancy is usually linked to the skill base of the team and to their contacts in the wider community, and especially those of the founder or managing director. For instance, this is the case in my consultancy. My first few projects were directly for people I knew or organisations I already had close links with. I love art and drawing and I have been commissioned to illustrate information leaflets on ecological matters several times – my ecological expertise could be relied on to get the drawing and associated information right. I also love teaching so we offer many courses for budding ecologists.

The profession

So to re-iterate, we mainly give professional advice including advising on legislation and the legal implications of the proposed development and mitigation advice. This is the professional element of the job - being a full member of a professional body (in this case IEEM), we have a Code of Professional Conduct, which must be complied with. IEEM provide excellent guidance on a range of topics including survey methods, standards for report writing, professional conduct and responsibilities and even how much to charge. The Code of Professional Conduct is reassuring for clients and many jobs, especially for local authorities, require full membership as this implies compliance. IEEM takes complaints against members very seriously and they have a disciplinary procedure to ensure we all behave professionally and to the required standards.

As part of our full membership we also have to have insurance depending on the role we are in. In my consultancy we have three types of insurance and if you are working for a consultancy then these should also be in place. Professional Indemnity (PI) insurance covers us for clients suing us as a result of the advice we give. We also have public and employer's liability insurance to cover our site visits and the staff. Full members of IEEM may need PI insurance depending on the work they are carrying out. Although I have never had to make a claim it is reassuring to have the insurance in case something goes wrong. For instance your advice could have major implications for timing of works and especially where protected species are involved lengthy and expensive delays can ensue for clients. Luckily, this can mostly be avoided by working closely with clients to ensure that survey efforts are robust and that advice is timely and correct.

I hope all this is not putting you off considering a career in ecological consultancy. As a trainee ecologist you usually get to do the good bits – getting out and about and doing surveys. Your reports are likely to be closely scrutinised before they leave the office and so the ultimate responsibility for the advice will not be with you until you are fully trained.

I commend the career to you if you like a challenge and would like to be a professional ecologist. We can make a real difference to the wildlife that is affected by developments and we have considerable power, backed up by the law, to ensure that wildlife is looked after and site enhancement is implemented. The confidence you gain with every increase in your knowledge base and skill is very satisfying. Your salary is likely to be much higher than your conservation colleagues, and should easily exceed that of most academics and maybe in the future you could set up your own consultancy.

In the next chapter I will cover more about how ecological consultancy has evolved into a profession. If you have already decided that this is not the career for you read on anyway, I might be able to convince you otherwise.

∞∞∞∞∞

Chapter 2

∞∞∞∞

The Professionalization
of Ecology

Working with the 'bad guys'

Back in 2003 when I became an ecological
consultant many of my friends and colleagues in
conservation thought it was a bad move.
Consultants were viewed with deep suspicion,
often seen as beholden to their paying developer
clients. With developers themselves having
gained a bad reputation in conservation circles, it
was often thought that we are paid by them so
presumably we could be 'bought' to cover-up or
ignore conservation implications of development
projects. I don't know if any consultants really
behave in this way as, in reality, we have great
power to influence our clients and to provide
mitigation for habitat loss. Before wildlife

legislation, and ecological consultants to interpret the law and provide advice to developers, this would not have been a consideration. I ignored "only in it for the money" and similar comments and I am glad to say that this perception has radically changed. Ecological consultancy is now firmly established as a profession and we are an integral part of wildlife conservation and management before, during and after development projects. We are also often hired by wildlife trusts or major landowners or conservation organisations to carry out research and survey work as they know that we have a high level of expertise and professionalism.

Back then, and to a certain extent still today, developers and architects (our main clients) were seen as the bad guys who weren't worried about destroying wildlife and habitats as long as the construction job was finished. In reality, they are just regular guys trying to do a job – providing us with houses, shops, workplaces, schools and other much-needed infrastructure. These facilities have to go somewhere and in general modern legislation ensures they are not permitted anywhere that has much ecological value. In my experience, developers just need educating about wildlife (and their legal obligations) - then they can better understand the need for the mitigation that we request. Once the developers and architects get the message that they do need to consider wildlife on all their projects, from the early stages of planning, then we find them very amenable to implementing our professional ecological guidance.

Strategic planning

Strategic planning is a legislative process to prevent development that would impact on valuable wildlife habitats. Local councils identify building land within and around cities, towns and villages. Then public consultation ensures that the right areas have been chosen (including avoiding areas of high ecological value) - this plan is then set for 10 or so years. The presumption is that building will not be permitted outside these areas, although inevitably it is to some extent. Although development is generally only allowed in areas of low ecological value, each site may still have some ecological value (hardly anywhere is a complete ecological desert). It is our job as skilled ecologists to identify important habitats and species and ensure that any losses are compensated for and that the impact on wildlife is carefully assessed and managed.

Changes in legislation

The growth in new wildlife legislation, as well as the strengthening of existing legislation, is probably the most significant factor that has led to the professionalization of ecology. As consultants, we have to be fully aware of all wildlife legislation, be able to apply this knowledge and then explain it to our clients (and often the planners too). In addition, many of us have to train our junior colleagues, and, in my case, I teach courses for budding ecologists - understanding the central role of legislation is a major part of such courses.

There are some key pieces of legislation and major events that have led to this increase in legal protection for some wildlife that you really must be aware of:

The Wildlife and Countryside Act

The Wildlife and Countryside Act was only created in 1981, the year after I finished by midwife training, and in the intervening time has been amended many times (species listed in Schedules 5 and 8 (animals and plants respectively) are reviewed every 5 years) with species added as they decline and become endangered. This is the cornerstone of our wildlife legislation and in many ways the first comprehensive piece of wildlife legislation ever to be passed into British Law. Since it was passed in 1981 it has changed significantly and has incorporated additional clauses and levels of protection laid down by Europe (e.g. Habitats

Directive in 1994) and the UK government (e.g. Countryside and Rights of Way Act 2000 – England and Wales and Nature Conservation (Scotland) Act 2004). It is a good idea to get a copy from HMSO (Her Majesty's Stationery Office) for your bookshelf, this will be the basis of much of your advisory work. When I left school in the late 70s working with wildlife was certainly not a career option and this was partly due to the lack of legal protection for wildlife.

In the late 1980's and early 1990's global interest in wildlife was sparked by the alarming trends in species and habitat declines. Vast tracts of the Brazilian rainforest were being lost and many of our UK species were on the brink of extinction. The disappearance of 'one species a minute' became the war-cry of those concerned about the imminent collapse of ecosystems. Most people do not give a moment's thought to our close connection to, and need for, the natural world – for fresh water, pollinators and indeed oxygen itself! The world needed to know how important it is to keep these ecological services working for us. Politicians turned out to be the ultimate tool for implementing action in the end as they have it in their power to create legislation – the stick rather than the carrot seems to work far better on humanity for some reason.

Rio Earth Summit

The 1991 Rio Earth Summit was the pivotal point for wildlife on the global stage.   From that meeting of the nations of the world The Convention on Biological Diversity, an international agreement, was formulated and has directly led to a sweep of new legislation across the globe.

Europe created the Habitats Directive in 1992 and this was transposed into law in the UK as the Conservation (Natural Habitats, &c.) Regulations 1994 (The Habitats Regulations) providing protection for sites in the UK containing habitats and species of conservation concern, and the full protection of some species of European importance, whether inside a protected site or not.   In the UK since Rio, species and habitats of conservation importance were identified under the Biodiversity Action Plan (BAP) at a national, regional and local level and the legislation has provided protection for special sites, habitats and species.   Wildlife (more accurately protected species) is now a material consideration for all planning applications with many species fully protected under UK and European law. In practice this means you cannot disturb them or even harm their habitat. Species such as the otter, great crested newt, dormouse and all bats are included. It is essential that you become familiar with the species on this list as they are afforded full protection including their habitat. As an ecologist, these are the main species that you will be working with on a day-to-day basis. There are many valuable resources online where you can find out more - in Appendix 1 I have listed some useful information and web sites for you.

Other legal protection for wildlife

Other important conventions and agreements gave fame to various cities around Europe over the years - internationally important wetlands were designated under the Ramsar Convention in 1971 and migratory species under the Bonn Convention in 1979. The Convention on the Conservation of European Wildlife and Natural Habitats – the Bern Convention, was an agreement to protect wild plants and animals in their natural habitats, and was one of the pieces of legislation incorporated into the Wildlife and Countryside Act 1981 (as amended).

Other European Directives have also led to an increase in wildlife protection in the UK and these include, amongst many others, the Birds Directive 1979 which set out objectives to protect birds that naturally occur in Europe; the Water Framework Directive 2000 which concentrates on creating a better water environment with particular regard to ecology; and the Environmental Impact Assessment Directive, introduced in 1985 and amended in 1997 which looks at the impacts of projects on the environment.

Dozens of other pieces of legislation have also been added to the list since Rio including the Badgers Act 1992, Countryside and Rights of Way Act 2000, Natural Environment and Rural Communities Act 2006, and Hedgerow Regulations 1997, - that is just a sample of the many English laws (Scotland and Northern Ireland have their own legislative bodies and have made different laws in recent years).

All of these agreements, laws and conventions are things that we, as professional ecologists, have to be aware of or implement on a daily basis. We need to be able to interpret the law, apply it to our projects and explain it to our clients! This is, for me, one of the most intellectually challenging but immensely satisfying parts of the job. Once you build this in-depth knowledge, and you know you know it, you gain great confidence when writing your reports, carrying out surveys and dealing with your clients.

We have to take into account a vast array of other legislation as well, to a greater or lesser degree, not all of it directly relates to wildlife. Unfortunately we cannot plead ignorance, especially if the particular law may affect what you may want to suggest for mitigation for loss of habitat. See Appendix 1 for a fuller list and useful websites.

To get started become familiar with the Wildlife and Countryside Act 1981 (as amended) and the Habitats Regulations 2010. Get your own copies and in particular become familiar with the Schedules - these are usually lists of species that are protected.

I am not mentioning all these to put you off or to worry you, but this is a large part of why we can now call ourselves professionals and why it takes time to become a recognised ecological consultant. It is not an easy career choice but it is very rewarding. Through reading, attending training and working with more experienced colleagues you will eventually learn the legislation, but while you are starting up then you just need an awareness of the key pieces of legislation and how to apply them for your projects and surveys.

PPS9 and site enhancement

Before I leave legislation there are a couple of other things that have also happened. One is the Planning Policy Guidance 9 (PPG9) – biodiversity and geological conservation – now superseded by Planning Policy Statement 9 (PPS9). This was produced for planners and sets out government policies to be implemented through the planning system and states that development should "preserve, enhance and restore the diversity of England's wildlife and geology". The key word here is ENHANCE. This is where we can start to realise that if projects are on areas of high ecological value then there is no chance of enhancing it for ecology. Therefore, most developments are now on sites of low ecological value. We also now have a stick to ensure that wildlife features are not lost on sites and that enhancements can be put in to make the site better for wildlife after the development than it was before – however, this is rarely easy.

We need legislation, and it needs us!

The levels of protection for wildlife that exist today are a far cry from our shocking post-war destructive behaviour which wreaked havoc on the countryside and decimated most species and habitats in the UK. We should no longer call the Common Dormouse common!

With a combination of the Rio Earth Summit, the Habitats Regulations, the Wildlife and Countryside Act, PPS9 and the vast array of other legislation to ensure protection of wildlife, the role of the professional ecologist has become clearly defined. Who else is going to be able to get to grips with the swathes of wildlife legislation and use their ecological knowledge to apply it to projects on a case-by-case basis?

Who else could give this advice to ensure that the developers and architects (and even planners) are complying with the law and doing their utmost to deliver biodiversity gain as required by PPS9? Who else is going to ensure that they do not fall foul of the planners or even end up being prosecuted for a wildlife crime?

Even since I set up Acorn Ecology in 2003 there have been many changes in the legislation, all of which appear to have increased the demand for good consultant ecologists. The market has grown exponentially over these few short years. Since the professionalization of ecology many of our conservation buddies are now willing to join us on the dark side! We do, however, need to know our stuff. This refers not only to the law but how to apply our knowledge of how ecosystems work, how different species react to certain changes in their environment, and most vitally – how to identify things.

The Institute of Ecology and Environmental Management

Our professional body is the Institute of Ecology and Environmental Management (IEEM) - it was established in 1991 to represent and support ecologists and environmental managers in the UK, Ireland and abroad. At the time of writing it has around 4,000 members drawn from local authorities, government agencies, industry, environmental consultancy, teaching/research and NGOs (Non Government Organizations). IEEM's main role is to provide a range of services to develop competency and standards in ecology and environmental management. All members are subject to a Code of Professional Conduct and the admission criteria are based on experience and time spent working in ecology. For full membership you need to have 4 years relevant experience.

I recommend Graduate membership to start with and after a couple of years experience you will be able to become an associate and put AIEEM after your name. IEEM is pushing for all ecologists to be registered as members and to follow the Code of Professional Conduct. Membership of IEEM, especially full membership (MIEEM) is now becoming very widely recognised.

IEEM produce many policies and good-practice guidelines, on topics such as report-writing, ecological impact assessment and useful documents about survey techniques (see Appendix 1 and Bibliography). They run low-cost training courses, seminars and conferences that are relevant to ecologists across the country (they are run by members such as me). Further along the line you can also apply for Fellow status and for Chartered Environmentalist (CEnv) through the Society for the Environment, which has affiliated membership with IEEM

(IEEM is a constituent body of the Society for the Environment).

If your professional life will be more geared to the aquatic environment you may want to consider joining Chartered Institution of Water and Environmental Management (CIWEM). Other professional bodies include the Institute of Environmental Management and Assessment (IEMA) and the Landscape Institute (LI) if you will not be dealing purely with ecology.

Ecological consultancy is now firmly on the radar of new natural science graduates, possibly because of the potential for it to be fairly lucrative, but also because there are not many careers where you can directly use your biology and ecology degree in the workplace.

And it's fun!

∞∞∞∞

Chapter 3

∞∞∞∞

What is an ecological consultant?

I am sure that by now you are getting the idea that ecological consultants are professionals with a high degree of knowledge and expertise which is built up over a number of years. I have already briefly explained some of the things you might expect to get involved in when working in an ecological consultancy, now we're going to go into a bit more detail as you have hopefully decided that this sounds like the career for you.

A literal translation of ecological consultancy means a place where you can consult with ecologists or where ecologists can act as a consultant to give advice. Many ecological consultancies are small departments of much bigger multidisciplinary environmental consultancies or engineering practices. The latter in particular often have ecologists on the team as they require their services so often it is worth them having their own in-house ecologists. Environmental consultancies may also have pollution experts, architects, structural engineers, arboriculturalists, landscape architects and many other disciplines all of which contribute to big strategic projects. Ecological consultancies that only do ecology can often be called in by large environmental consultancies to carry out the ecology section of the scoping or mitigation works or contribute to the EIA (our bit is called the EcIA – Ecological Impact Assessment). Either way, there are job opportunities for you in all these types of businesses.

Types of project that will require an ecological consultant:

- Surveys of development sites for protected species, habitats and flora.
- Wildlife site assessments (e.g. SSSI assessments, County Wildlife Site assessments).
- Management advice for landowners (e.g. quarries, nature conservation landowners (e.g. National Trust, wildlife trusts)).
- Management plans for ongoing management of sites.
- Habitat creation and enhancement schemes.
- Ecological impact assessments (e.g. wind farms, large developments, road schemes) as part of the environmental impact assessment.

- Applying for licences to disturb protected species.
- Translocation of species (e.g. reptiles, great crested newts).
- Preparing mitigation plans or strategies for developers where protected species are present on development sites or where sites need to be enhanced for wildlife (PPG9).
- Giving evidence at Public Inquiries as an expert witness.
- BREEAM or Code for Sustainable Homes assessments.
- Invasive species eradication plans.
- Specialist surveys (e.g. botanical, bats, invertebrates)

Project life cycle

Projects come in all shapes and sizes. They can require surveys in rural or built up areas, night work (e.g. bat surveys), early starts (e.g. dawn surveys), meetings with clients and other team members, or working with other consultancies on large collaborative projects.

A typical project, carried out by a single consultancy working with a developer often goes like this:

1. The client gets in touch and explains the situation and what they require for their planning permission.

2. We discuss with them what we need to do, how and when. We then prepare a quote.
3. They commission us and arrange for access and provide us with other information such as site maps and proposed development maps.
4. We carry out a scoping survey at the site to assess the likelihood of protected species and to assess the site for important wildlife features.
5. Once we have carried out the survey we prepare a report of our findings, conclusions, suggestions for further survey and mitigation suggestions. We invoice the client.
6. This report will then either lead to further survey or be submitted with the planning application.
7. If further surveys are required these are discussed with the client, a quote prepared, and plans made to carry them out — depending on the type of survey they can be very time-consuming, requiring surveying over a number of weeks, months or even years for very large projects.
8. If protected species (especially European protected species) are on site we may need to provide a detailed mitigation plan to the planners as part of the planning permission. We charge the client for this.
9. If a European protected species is present then we prepare a European Protected Species Licence and liaise with the client and Natural England (or equivalent) to get the licence. We may then need to attend to supervise e.g. roof or hedge removals, or pond dredging. We also need to report on the actions taken during the works. We charge the client for each of these stages.

The entire project may involve a range of activities and skills which make up our role as consultant ecologists. When you start out in your ecological career you will only get involved in limited parts of this process. As you progress you will take on more responsibility for the whole service supplied to the client.

Trainees at work – your first job

As a trainee you are likely to get the time-consuming surveys such as reptile surveys and bat activity surveys (after training of course). If you have a problem with reptiles – snakes in particular – you may need to get yourself de-sensitised by just doing it and getting experience, or having a re-think. Could you catch an adder? I run a reptile surveying and handling course that has cured many of a snake phobia – it's amazing how quickly something scary becomes routine after you've done it a few times. These surveys often require basic skills and don't require licences but give you the opportunity to learn a new skill, whilst doing something useful and generating income for the practice. For the senior staff it means that we can concentrate on more complex tasks or projects.

Hopefully you will also be given the opportunity to go out with more senior consultants and will quickly pick up information and experience that will stand you in good stead for developing your career. Obviously if you are interested, enthusiastic and willing to put in some time to do background reading you will get a lot more from this experience than if you just do the minimum and don't ask your colleagues questions. Don't forget that you MUST spend as much time as possible developing your identification skills – colleagues can help but in the end this information has to go into your head and only you can put it there. This is where some personal effort and dedication will pay real dividends.

You will often be taken out on surveys just for health and safety cover. For example bat work at night, working around water, or maybe to help carry equipment. This is an excellent time to chat to your colleagues and find out about how they developed their career, ask any questions about the survey method, what you are finding, what they might advise and any other burning questions you have. If you are working for a big consultancy you may be required to travel a lot too, again a good time to talk to a colleague. I find trainees like to discuss the legal aspects of our findings and what advice we will give. Getting to grips with the legal side of our advice is one of the most difficult things to master and takes time. Trainees also like to ask about how I set up my business.

We have developed templates for our survey

reports and this ensures a consistent product and also saves time. In your first job you may be asked to complete simple reports at first and gradually do more and more complicated ones. Good drawing skills are useful as most reports have some sort of sketch map or you may need diagrams of mitigation suggestions. My new trainees often start by doing the map and preparing a results table to go into the report. Some ability to use a digital camera, download photos and re-size them helps as most reports also include photos.

As you gain more experience, under the guidance of your senior colleagues, and backed up with your own studies, you will soon be able to tackle more varied tasks. Remember your senior staff should want you to gain as much experience as possible – you will then be far more useful!

You might find it useful to read IEEM's student documents 'Ecologist Profile' and 'Rooting for a Career in Ecology or Environmental Management?' and 'Field Ecology Skills Guide' available on their website - www.ieem.net.

Summer madness

The summer is the busiest time in the ecologist's year. After all, the bats, reptiles and dormice are active, the flowers are out and the insects are buzzing. This is great if you like being outdoors - you get to work outside in lovely warm sunny weather (OK, sometimes it is not sunny. You might even get soaked to the skin). Unfortunately, the summer also means the days are really long and, particularly if you are into bats as well as everything else, you might end up doing ridiculous hours. But it should be fun. Hopefully you will get the time back when the season quietens down a bit. You are unlikely to get paid overtime and will often be expected to do whatever it takes to get surveys finished on time and the reports written up. For this reason be prepared for long tiring summers and fairly boring winters. My solution to this is to have my holidays in the winter – that way I get sun all year!

Much of the work in the summer is as a result of an initial survey on a development site that has identified potential for certain species (e.g. long grass may harbour reptiles, a pond may have amphibians). The initial survey can also include mapping the habitats and it will dictate if further survey is needed. Further survey may have to be delayed until the following survey season, especially if you carried out the initial survey in say October and the survey season starts in late April or early May. This can be the source of some distress to less experienced developers but in the end they have to accept the situation. Often they will still submit their planning application without the further survey and the planners will throw it out until the further surveys have been completed.

By law protected species are a 'material consideration' for planning and the planning authorities need evidence that none are present on site before granting planning permission. At this point we are often accused of making work for ourselves by our clients. But we know that the SNCOs and the planners will ask us why we did not ask for further survey if there is potential for protected species. It would be unprofessional of us to gloss over our findings and that would never benefit wildlife, which to me is our ultimate aim.

Planners do sometimes grant planning permission without further survey and make it a condition of planning that a survey is carried out before any work take place – technically this is illegal. I speak to planners that allow this and point out that we may find something really important and then the plans may have to be altered to provide mitigation, or the SNCO may override planning permission if certain conditions have not been met – all very tricky, and issues you will get more confidence in as your experience grows.

There is no typical day or week

As our work is seasonal, very varied and reliant on our customers appointing us, no day or week is the same as any other. If you work in one or two regional counties your working day will often consist of surveys in the morning and reports in the afternoon and bats in the evening (hopefully not every evening). If you work for a consultancy that works all over the country then a large part of your day could involve travelling and you may have to stay away overnight for several days. Your working year will probably consist of manic summers and slow winters. Your working week however, will almost certainly be very varied and interesting.

We deal with a variety of other things including preparing information, giving advice, preparing management plans, giving mitigation advice and implementing it, species monitoring post development, research and specialist surveys, and preparing licence applications for disturbing protected species. As I outlined in a previous chapter, you can be asked to do almost anything to do with wildlife by your clients. You will probably not just be working with developers - we regularly carry out work for local authorities, government organisations, wildlife trusts and more.

This is one of the most interesting, challenging and intellectual careers for a natural sciences graduate available. I expect you are now wondering how you can prepare yourself for this career and what skills you should be trying to acquire. The next few chapters cover the skills and knowledge you will need, how to get experience, how to prepare a CV, and how to succeed at interview and get that first job. We will also briefly look at your continuing professional development.

Consultancies are businesses

All consultancies are run as businesses, even the ones that are under the umbrella of a local Wildlife Trust, it is just that they give their profit to the Trust and not the shareholders. Businesses are there to make profit, if they don't they are not viable and will not survive. With this in mind, especially with smaller consultancies, you may get involved in some aspects of running the business such as admin, quotes, accounts, marketing or following-up on leads. Although you may think this is not as interesting as the ecology, most of my staff find it an excellent way to gain other transferrable skills and, maybe, prepare them for the future if they want to set up their own consultancy.

You will always have to record your time and mileage for invoicing and forgetting to do this would be a serious matter – after all you don't want to be spending time on something if you are not being paid for it. Getting to know the systems of the business can take some time but are worth knowing as they are designed to make the business more efficient. In smaller consultancies you will always be welcome to suggest ways that systems could be improved. You must always bear in mind that your attitude and conduct at work reflect onto the business (and ultimately your job). A curt manner on the phone or rudeness to a client will result in a loss of work and this ultimately leads to less money and the possibility of losing your job. A professional, positive, helpful manner is always worth nurturing. Incidentally, how you conduct yourself is important at interview, especially if going for a trainee job where you lack experience, as your attitude may be the only factor which will get you the job. More on that in Chapter 8.

ᴏᴏᴏᴏᴏ

∞∞∞∞

Chapter 4

Applying knowledge, expanding skills

You have very likely got, or are still studying for, a BSc degree in biology, zoology, ecology, environmental science or something similar. In that case you have a bit of a head start. People without degrees are usually at a disadvantage unless they know people in the business who can give them a break and they already have really excellent field skills or lots of experience. Degrees in other disciplines such as geography can find a niche too but you would have to have some relevant element in your degree or have gained extensive ecological and field skills experience to compete with other job candidates who are more biological. I know of one recent graduate of geography who was actually known as a top birder and so his ecological credibility was already in place from his own personal experience and activities outside university.

From the broad range of biological modules on offer in your second and final year as an undergraduate I think it is useful to gain a good understanding of genetics (inbreeding, gene flow, evolutionary processes), metapopulations, effects of habitat fragmentation, niche, viable population sizes, animal and plant physiology and taxonomy. A biological element to your degree should ensure you have this.

Undergraduate skills

Whilst at University you have been acquiring knowledge and skills, some of which you may not even be aware of. I went to University at the age of 39 so I didn't have the whole issue of having to cope with leaving home, managing my time, my work, my home, and meagre money for the first time or establishing a social life. These were already in place, but these are all skills in themselves that young undergraduates have to acquire on top of their degrees. In fact my daughter, aged 13, stated that I was only coping with my degree, family and housework because I had no social life! My social life was obviously not in the same league as my younger fellow students but I had spent plenty of time socialising many years before. Many of my younger fellow students complained that it was easy for me as I didn't have all the social pressures they had. Obviously my maturity gave me the ability to focus and get a lot done in a short time - I had developed my own resilience to stress and pressure. Perhaps they would have been more sympathetic if they had a home and children to raise!

Other skills you develop in your studies are writing skills and putting forward an argument or some piece of information derived from your research. Your familiarity with scientific papers, referencing, proof of evidence, robust data, methods, results, conclusions and so on, all hold you in excellent stead for writing reports of your findings as a professional ecologist. IT skills are also pretty essential in consultancy and University courses all of which include an element of IT. Although presentation skills are not often required they are a confidence-booster - you may well be called on to speak at a conference or seminar at some stage, run a course, speak up at a big meeting or give evidence in court.

The most noticeable thing I found with doing a science degree was that it changed my mindset into one of a scientist. I was no longer able to accept sweeping statements unless they had proof or I could think through logically whether something could be the case. Having a scientific approach to your work is essential. Ecologists are, after all, applied scientists.

Postgraduate degrees

Many people finish their first degree and start to realise that they don't actually have any of the skills they need to get a job. The next logical step can be a Master's degree (MSc), which will further develop your knowledge and skills in one area. A word of caution though - make certain that you are choosing the right course. Find out what courses other people have taken who have got jobs in consultancy and what they thought of it, and find out what skills you will end up with if you do the course. Does the course offer the right skills in the right quantities to get you the job you are after? Most MSc courses are one year full-time or maybe two years part-time. They can be expensive and if you are studying full-time for another year then you will not be able to earn much. If you are really struggling to get a job then an MSc is probably a good move but, as I describe in this book, there are other ways you can get skills and knowledge and those might be more flexible and tailored to your needs (e.g. volunteering, short courses, work experience). If you can find a course that you think will turn you from a biologist to an ecological consultant or environmental manager and you are struggling to get other relevant experience then it may be your best option. In the end, you are the only one that can decide what is best for you, and the time and the funding you have available. I am fairly confident that if you managed to get a job in a consultancy for a year you would be slightly better off than someone who spent that year doing an MSc. Equally, going on to do a PhD may not get you that job immediately but with some experience in consultancy you would certainly have more kudos in the long run.

Whilst on the subject of master's degrees, I studied an excellent (but very expensive) one at Bristol which gave me all the skills I needed to get going straight away and I am sure there are many other suitable courses across the country that you could attend. You do not necessarily need to have a degree either. Several of my fellow students had come from conservation backgrounds and had extensive experience in their fields. Make sure the course covers the topics outlined above as well as Environmental/Ecological Impact Assessment and a good dose of field skills. Full-time courses obviously mean that you have another whole year of study which is expensive in terms of fees and lost earnings. Mine was part-time over 2 years so I could earn some money in the meantime. Short courses, as an alternative, can be fitted into your own time scale, so you could fast-track yourself and get them done in 6 months or take a couple of years whilst spending spare time consolidating your learning or volunteering to get experience. If you need to work full-time, raise a family or have other commitments then short courses are probably the only way you can fit it all in.

Identification skills – your career starts here!

One thing that many degree courses conspicuously lack is field skills, in particular, identification and taxonomic skills. We could blame the universities to some extent - they have increasingly withdrawn from fieldwork back into the lecture theatre. However, in my experience the field skills, especially species identification, are something you have to do for yourself. No one else can make you remember 300 plant species for instance. It is up to YOU to get this stuff into YOUR head. This task is somewhat easier if you have a good tutor and good fieldwork opportunities but your own interest and attitude to this is the key to gaining good ID skills.

My bedtime reading for at least a year was to flick through the Collins Field Guide to Plants – looking at the pictures, finding plants I had seen in the field, going though any I had collected or photographed to identify them, going through keys, looking at plant descriptions, family descriptions and plant community lists for each type of habitat. After a while I got pretty good – but it takes a lot of practice. Likewise if I wanted to know all about bats, dormice, badgers, otters, dragonflies, butterflies, or whatever, I got the right books and read up about them. My favourite books are the Whittet books which give a good introduction to a range of animals and have great cartoons and drawings which appealed to my visual learning style. Don't limit yourself to the academic reading lists you need for your degree.

Most students also forget that you have at least 5 months off a year to get plenty of other reading and field studies done! This time is valuable for your future, it is your chance of optimise your learning whilst still at University, maybe to have a look at things that interested you in the course and you didn't have time to find out more, an opportunity missed by most students. You are unlikely to have that much free time again until retirement. Later I will be making some more suggestions as to what you could be doing in your holidays to get ahead in this career.

Make the most of your time at University (and your free time), the knowledge you gain here can be applied in the workplace and a logical, methodical, scientific approach makes your work much more professional.

Botanical ID, is the most lacking skill in trainee ecologists, whether you have a degree or not. Most consultancies lament the problems they have finding trainees with even a basic grasp of plant ID. How, for instance, do you expect to describe a woodland in a survey report if you don't even know if you were standing in an oak wood or an ash wood?

Sorry to keep going on about this but it is important. Gaining ID skills is overlooked by most students, or aspiring ecologists, but remains one of the most important skills needed by ecologists. **Honing your ID skills is the number one way to get ahead in your professional ecology career**.

If you are not sure where to start then a Beginner's Botany course is a great way to start (I run one through my company and through IEEM). With an inspiring tutor you will find that you at least know where to start, what to look for, how to use keys and how to identify common plants and trees. I recommend beginners to take out with them the Field Studies Council laminated guides to plants. There is one for each habitat and you are then only looking for a limited number of plants, maybe by flower colour, rather than having to find your plant in a large ID book with 3000 species. A recommended reading list is included in the bibliography – there is also a regularly updated one on the IEEM's Survey Methods web pages (**www.ieem.net/**surveymethods).

Your best route to meeting local botanists who may be running courses or guided walks locally, is to join the local wildlife trust (see a list of these at www.wildlifetrusts.org) or join a botanical organisation such as the Botanical Society of the British Isles (www.bsbi.org.uk). You should also look out for tree walks by the Forestry Commission (www.forestry.gov.uk). Many wildlife trusts hold training days on a variety of subjects that are quite cheap to attend.

For the many other groups such as mammals (including bats), birds, amphibians and reptiles, moths, and dragonflies you will find local experts, guided walks, surveys, talks and other events going on locally. All of these are useful to tap into. They are also a great way to meet new people. Don't limit yourself to the University; there is a wealth of other opportunities out there. In the next chapter I will be covering how you can use volunteering as a valuable way of gaining skills and experience.

Courses

Don't limit yourself to the courses and modules you will do as part of your degree. Expand your knowledge base where possible with as much relevant information as you can - don't forget your 5 months off a year! For those without a degree, or a non-relevant degree, short courses are an excellent way to get up to speed with what will be required for your new career.

There are some essential skills that will propel you into an ecological consultancy job much more quickly and courses are often the way to get these skills. I see many of the short courses that I run as ways to unlock the 'trade secrets' quickly. It is very valuable to have a tutor available to you who is experienced, knowledgeable and can direct you not only to survey techniques, specific species ecology or mitigation techniques and answer any questions you have, but also guide you to background reading and give you real live field experience. This is an excellent way to spend a day or two.

I recommend you try to attend course covering the following skills if you want to become an ecological consultant (if your university degree does not do these already):

- Phase 1 habitat surveying
- Reptile surveying
- Protected species surveying e.g dormice, badgers, bats, great crested newts
- Site assessment and report-writing
- Beginner's botany
- Wildlife legislation
- Invasive species

There is no standardised training for ecologists and everyone comes from different backgrounds and experiences but at the end of the day there are some essential skills that we all need and we all need to have a consistent approach to our work to remain professional. Most trainees will be expected to have basic survey and identification skills and a good understanding of a range of habitats. The fields you are most likely to cover at work are often geographically limited, for instance we have very few great crested newts in Devon but all species of bats, and lots of dormice, whilst in other areas you may do lots of crested newt work and nothing on dormice.

Apart from my company and IEEM there are very few organisations or universities that offer a comprehensive range of short courses for aspiring ecologists. It is really a case of hunting around for local opportunities as well as consolidating your skills with specific formal courses. The Field Studies Council (www.field-studies-council.org) is starting to introduce some more professional ecological courses - they mainly provide an excellent range of identification courses that I highly recommend. The Mammal Society (www.mammal.org.uk) also runs excellent courses on surveying and identifying mammals. The Natural History Museum does Identification Qualifications in various taxanomic groups. University of Birmingham, in collaboration with the Field Studies Council and the Botanical Society of the British Isles (BSBI) offers a range of accredited courses in biological recording and the University of Leicester offers a similar modular course leading to a Certificate of Higher Education in Plant Studies and Field Botany. Many other organisations also run courses so keep your eyes open.

Other skills you'll need to develop

At this point I need to cover some of the other skills I think you need to become an ecologist. This is the *je ne sais quoi* that is hard to define about what it takes to survive in this career.

According to IEEM these are the qualities that make a good ecologist or environmental manager:

There are general qualities, such as self-motivation, teamwork, computer literacy and communicating and negotiating skills, which are necessary for most areas of work, but special qualities are needed for success in ecological work. These extra skills include:

- a fascination for animals and plants;
- a thorough knowledge of the functioning of natural systems;
- good academic qualifications in biological or environmental subjects;
- expertise in one or more groups of living organisms;
- the facility to infect others with enthusiasm about the natural world;
- enjoyment of fieldwork;
- the staying power needed to complete tedious and sometimes uncomfortable tasks in field or laboratory; and
- an objective approach to conservation issues.

One of the top things, which IEEM only hints at, is fitness. This is 'not an easy way to earn a living' my elderly father declared once when I took him out for a day on remote fieldwork on Dartmoor. The day was long, strenuous and hot; luckily I knew he was at least as fit as me. In this job we often climb up and down hills, walk for miles, carry equipment, go out in all weathers at all times of the year, go out at all hours, and all whilst concentrating on collecting data. We often work very long hours if we do night work such as bat or newt surveys or dawn bat or bird surveys. We get tired, thirsty, hungry and dirty. Fieldwork is tiring, particularly if you do bat work – a working day may be from early in the morning until the early hours the next day and may also involve extensive travelling. If you are not fit you simply will not be able to cope with the fieldwork. Fitness is something that you can work on, much like identification skills, and is also something you will gain from doing the job!

If you get a job working for a larger consultancy you may find yourself travelling around the country much of the time depending on where their clients or projects are. I have known consultants drive for 4 hours to carry out a one-hour survey and drive for another 4 hours to get back. Personally this is something that I would hate and so I stay in my local region. But you may be working in Kent one day and Durham the next. You may spend several nights away from home each week too. Find out about the travel distances when you go for jobs because this might be something that you don't want to spend most of your time doing. You obviously need your driver's licence and a car as most places we survey are not on bus routes and being a fairly confident driver is very useful. We often drive down very narrow lanes or even go off road. You are unlikely to be provided with a company car so find out what the mileage rates are when applying for jobs to ensure you are adequately recompensed.

If you love being outdoors in the countryside then you will be well suited to this career. It may be boiling hot, pouring with rain or a blizzard but we still go out. If you cannot visualise yourself in these situations then you need to seriously reconsider whether professional ecology is the right career for you. It is not easy coping with the conditions we sometimes find ourselves in (after all it might have been lovely weather when you set out). We have a saying 'there is no such thing as bad weather, just inappropriate clothing', which is true to some extent but you do need to have an element of resilience and know your limitations.

The first full-time trainee that I took on had a great CV, as did most of the others who applied. She listed her hobbies as: hill walking, climbing, camping and surfing – amply demonstrating to me she was an 'outdoorsy' person. I have noticed those listing badminton, knitting and tai chi have proved to be less than able to cope with the physical side of the fieldwork.

Good people skills are essential as you may be dealing with clients who are not happy about what you have found; irate neighbours who don't want a development next door; architects, planners and colleagues. New clients phone up out of the blue and you will need to build an instant trust and rapport. You may be tired from an exhausting week but still need to deal with people calmly, confidently and professionally.

Many of you will have a natural ability to handle other people and have no problem with this aspect of the job. Others may be shy, lacking confidence or self esteem and find this difficult. If you have had jobs where you have had to deal with the public, such as shop or bar work or any customer interaction this should stand you in good stead - confidence will come with experience.

Finally, one of the most important qualities you'll need to develop is a healthy dose of bravery. A willingness to have a go at something you might be extremely uncomfortable doing. A willingness to do things that others may think are a bit dangerous. Ask yourself how comfortable you would be doing some of the things we could be asked to do – how about crawling around in a dark, confined, dusty, spider-ridden loft or dark damp cave looking for bats? How comfortable would you be going up a 15m ladder to check a nest box or to gain access to a loft in a large building? Could you pick up a bat or a mouse or an adder? Are you afraid of the dark?

As you can see, these other skills are all part of the make-up of an ecological consultant. If any of these things fill you with terror then beware! You could be asked to do this at any time and you need to be prepared mentally and physically.

We are a fit and intrepid bunch with great knowledge of wildlife. We have the ability to apply our knowledge and to be professional at all times whilst successfully handling people from all walks of life. If that describes you then you've found the perfect job! Personally I love a bit of excitement at work, I fear little (except maybe accounting and large horses) and will have a go at anything.

∞∞∞∞

Chapter 5

∞∞∞∞

Volunteering

Volunteering is often the only way you will get enough experience to make a start in this career. It does seem to be part of the rite of passage for getting into ecology and conservation. I have certainly spent massive amounts of my time on it over the years – as a volunteer you provide a great resource to NGOs, gain huge amounts of enjoyment from helping worthwhile causes and pick up essential skills. Although needing to do a lot of volunteering can seem unfair, especially if you are cash-strapped after your degree, if you use volunteering for *your* benefit as well as the NGO's, then it will make an enormous difference to how quickly you can progress. I still volunteer and I am sure that almost all other ecologists do also.

My volunteering

Last year I spent one evening a week during the

summer trapping bats in woodlands around Devon as a volunteer. I have also carried out a couple of free bat walks and children's events. How could that possibly be an advantage to me career-wise? Well for the bat trapping I spent a long weekend on training to learn harp trapping with one of the best bat workers in the country - an enormous privilege for me. My bat ID and field skills improved enormously during the course and the surveys over the summer. I spent quality time with some amazing people who have given me leads to new work and opportunities. It was also an amazing experience to sit in dozens of woodlands across Devon at night. I've picked up new knowledge about my local area and increased the records of bats in the county. This in turn has made me a better teacher for my students on the bat courses I run.

Volunteering can sometimes be the only way that you can get experience and something tangible on your CV. If organisations are looking for volunteers they will assume no knowledge so you will receive full relevant training and learn new skills. You will also become known in the ecology and conservation world and make useful contacts. The more you get involved the more you will gain great skills and grow your professional network.

One of the first volunteering survey jobs I did was in Devon about 10 years ago. Operation Otter sounded intriguing - Devon Wildlife Trust wanted to know where otters were in Devon and the Trust did not have enough manpower to check all the rivers in the county in a short period of time. To make up for the lack of staff they trained volunteers to spot otter field signs and to assess the river habitats. Then we all went out in one weekend to give a snapshot of the status of otters in the county. You can immediately see that this project not only had advantages for the Trust but also for the volunteers. We not only contributed time to help the Trust, and of course the otters, but gained a new skill and something else for the CV. From that survey it was found that otters occurred on every river catchment in Devon. Having suffered massive declines over the 60s and 70s, this was an amazing discovery! It was a great feeling to be part of that finding. From that point on the conservation of otters in Devon could be based on a solid baseline of survey data.

I have found volunteering particularly valuable in gaining skills and I will give some more examples shortly. I have found, however, that your volunteering really needs to be focused on *you* and *the results you want to achieve*. There are many opportunities out there but be careful. If you spend every weekend scrub bashing you may have fun but what will you learn? How will that help your career (remember the three different paths?). If you want to use your volunteering time to become an ecologist you must focus on survey or field skills primarily. Your time is precious and so you must be sure that the volunteering you do will give *you* something, as well as the organisation which is offering the work.

Other volunteers

A student of mine impressed me (at first) by saying that she had volunteered for two days a week for her local wildlife trust for a year. She was trying to get into ecology and conservation. I asked what she had been doing: "reception, photocopying and envelope stuffing" she said. She was, not surprisingly, feeling a bit frustrated that her volunteering was not getting her anywhere. If she had the sole (and very much worthwhile) objective of helping the wildlife trust, for example, if she was retired, then this role would have been fine. But she wanted to gain something for her CV. I urged her to stop that role immediately and at least start volunteering on the reserves. I also told her about other survey volunteering she could carry out. A year later the change in her confidence and demeanour was remarkable. She spoke with authority on the many topics she now had experience of and she had also been studying hard for her degree. When she finished her degree she was already armed with knowledge *and* skills and had made many valuable contacts in the local conservation and wildlife community which would stand her in good stead in the future.

My daughter got the ecology bug when she was about 15 or 16. I think she saw that I was having fun with my career and she loved wildlife too, so it made sense to follow this as a career as well. We sat down and looked at her interests and started to formulate a plan to get her some experience by volunteering and getting involved. I knew it was not enough just to have a degree on your CV, especially as she would be competing with dozens of others who also had a degree. I also knew that it takes a long time to build a good CV and that she should start straight away.

My daughter's interest was mainly in birds, so we worked with that. The first thing we did was volunteer at the local RSPB reserve together as a weekend warden - we did this for just over a year. That gave her the confidence to go off and do her own volunteering. We then went on the local British Trust for Ornithology (BTO – www.bto.org.uk) training day for field surveyors for the national breeding bird survey. We carried out nightjar and tawny owl surveys, breeding bird surveys and also wintering bird surveys as weekend wardens. A friend of hers had won Young Birder of the Year and she decided she could do that, but she had to complete it before she was 18 or that chance would not be available again. She spent hours every weekend and many mornings before she went to college, studying the birds of our local area. I taught her habitat mapping so she mapped the whole area.

She wrote up her observations and submitted it to the competition judges (she did not even let me read it before it went off!). To our great joy she was awarded Young Birder of The Year with that very project and we all went up to the Bird Fair at Rutland Water to collect the prize.

The following two summers she spent 8 weeks volunteering on RSPB reserves where she gained a variety of skills including surveys, habitat management and guided walks and got to spend quality time on a well-managed nature reserve where she could indulge her passion for birds every day. She also had the added advantage of working with me for my ecological business but mainly she was gaining her own experience. This experience would not only stand her in good stead for later life, it also secured her an unconditional offer at the university of her choice. When she went to University she got involved in the local wildlife trust and RSPB groups, carried on birding to keep her skills up and in the holidays did some volunteering with RSPB. She also managed to secure her first paid job whilst still at university.

You can imagine, therefore, that by the time she had finished her degree she had a raft of skills and experience that instantly put her head and shoulders above her competitors. The first job she went for she was offered but she decided it would not take her in the right direction so she turned it down. The next one, that she really wanted, for which they had 90 applicants, she got the job and accepted! She was obviously lucky to have a professional ecologist parent, but it was her years of quality volunteer experience that got her the job in the end.

Voluntary work placements

In some companies, mine included, there are voluntary work placement for graduates. In my company this is a 4 week placement where the volunteer works for free. Many post-graduate degrees offer work placements as part of the course. Students and volunteers learn many new field skills, report-writing, legislation, ID and many relevant skills, including an insight into running a business. We try to tailor the placement to the interests and needs of the individual. The aim is to make them feel like they have gained an enormous amount from the placement. It is of mutual benefit too as we get an extra pair of hands at our busiest time of the year. So far there has been an extremely high number of them getting jobs almost straight away.

For this placement we are aiming for graduates with no or very little practical experience, although we have had volunteers that are birders and wanted to get other experience, and others that have worked abroad but lacked UK experience. Each summer we have 10-12 placements. We gain a lot from having enthusiastic new colleagues and they gain enormously from learning new skills and having lots more to put on the CV. If I am hiring then I will also know which ones I want to hire.

I am not sure if one of your local consultancies does something similar but I have found that 4 weeks is not too long if you need to fund accommodation, but it is long enough to get some really valuable experience. Offering to work for nothing at your local ecological consultancy is always worth a try. Make sure it will be to your advantage though, and that you will get a good variety of real experience under your belt, you don't just want to be making tea, adding names to the database or filing.

Making volunteering work for you

Hopefully by now you can see that working for free is a valuable and essential part of getting a start in this profession. And you can often do it all in your spare time.

Before you start to volunteer you need to be clear what *you* want out of the experience. Is it a new survey skill? Is it an increase in knowledge? Is it new contacts? Is it an insight into how reserves are managed? Once you decide on the skills you need then you need to be sure the volunteering will deliver that for you. What are they offering you? What will you get from the volunteering that will help you in your career? What contacts will you make? Can you put new skills on your CV? How relevant is the new skill you will gain to your career aspirations?

Look out for opportunities to carry out surveys. Sometimes they are not quite what you are looking for but if they are at least moving you in the right direction they are worth going for. Many are for just a few days a year so you need to keep looking to make the best use of your time. My local wildlife trust has a volunteers' newsletter, so does the local RSPB and BTO. Each give lists of projects that they need help with. Then just volunteer - they will welcome you with open arms!

Create your own voluntary work placement

You can also invent your own volunteering jobs. For instance, I noticed that a local site was becoming scrubbed up. When I enquired about the management to the landowner they said no management plan had been prepared yet and they were not sure what to do with it. I offered to survey the site, map the habitats, prepare a management plan and talk to them about my findings. I then encouraged them to manage the site for the benefit of wildlife. I also offered on-going monitoring of species and changes in habitats with the management plan in place. At the time I was studying my master's degree and I needed to put some of my knowledge into practice. This was perfect! It enabled me to consolidate my learning and give me some tangible experience. I was unable to commit to ongoing support at the reserve but offered to train up the next volunteer and so the reserve is now monitored and managed for the benefit of the wildlife as an ongoing process.

Where to look for good volunteering

I have already covered some ideas but here are some more good opportunities to research. Hopefully a few of them will help inspire you to get out there and volunteer.

NARRS is a very good example, it is the National Amphibian and Reptile Recording Scheme (www.narrs.org.uk) where volunteers are taught standard survey techniques and given a specific area to study. Not only does the volunteer gain useful training and experience but the knowledge of our reptiles and amphibians is enormously expanded so that conservation efforts can be more focused. Organisations that provide such training are actually leveraging their time by training dozens of new people to help them. Again, no knowledge is assumed and the training gives in-depth knowledge of the species and the survey techniques. Once you have carried out your survey you also have experience.

If you are interested in bats then there are many opportunities to help out. Firstly try your local bat group. Some are very active and others not so. Try to find a bat worker to tag along with to get experience. Eventually you can work towards a bat licence that allows you to enter bat roosts and give advice - this takes time and you need a fair bit of experience, so the sooner you start building up your bat knowledge and experience the better. The Bat Conservation Trust (www.bct.org.uk) also carries out a range of national surveys that do not require a licence. They explain the survey methodology and you just go out and do it at the allotted time. Several different species are covered by different methods; all are interesting and beneficial towards getting you your licence.

Your local wildlife trust is usually looking for volunteers too. In my experience they are a little slow at getting in touch with people who offer to volunteer. Maybe a good starting point is to go for a day activity where you just need to turn up. They manage reserves that need management and monitoring so there is plenty of scope for survey work – and don't forget my story above where you can create your own volunteer job.

They often have county-wide projects such as the Operation Otter project I have just described where you can receive some very useful specific training.

I have already mentioned the BTO – they hold very good training days, although I suspect that some areas have more than others. RSPB does excellent residential volunteer placement across the country for periods of 1 week to a year. You could, for example, help protect nesting ospreys in Scotland or monitor swallowtail butterflies and marsh harriers in Norfolk. They often provide accommodation and training, this is an excellent experience to work on different reserves learning about the species and habitats there. The RSPB also has many other volunteer opportunities that you can fit into your spare time. Make yourself known at your local reserve or regional office to find out about opportunities.

Butterfly Conservation (www.butterfly-conservation.org) has volunteering opportunities. The local branches may offer training for certain species or projects. Nationally there are a range of surveys where they need surveyors; some have training and others have excellent documents telling you how to carry out the surveys and identify different species so you can study your own local area. Local groups often hold events, seminars and talks. Nationally there are good conferences too.

Mammals are covered by The Mammal Society

(TMS - www.mammal.org.uk) and the People's Trust for Endangered Species (www.ptes.org - previously known as the Mammals Trust UK). TMS carries out many national monitoring schemes as well as brilliant training. PTES funds projects and also co-ordinates volunteer projects such as the Great Nut Hunt – a national survey project for dormice.

These are just a few examples of where to start. Why not ask other people who work in conservation what training or surveys they are involved in? Ask to be kept up to date with events by subscribing to newsletters. Watch web sites of your favourite organisations for ideas on where to help out. A list of useful websites is given at the end of this book in Appendix 1.

You are inspired, when should you start?

As you can see there are many opportunities out there and once you grow your contact network you will hear of other events and surveys. When should you start? NOW of course! If you have yet to complete or even start a degree then still start now. If you are in the middle of a degree, start now. If you already have a degree, start now. If you want to change your career, start now. If you already have a job but still want new skills, start now!

Volunteering is a particularly useful way to gain skills if you are working too, as you can fit it into your spare time. You are going to gain skills, meet like-minded people, get known, show people how good/nice/keen you are and increase your wildlife confidence. I have found that knowing lots of people is always useful as they can help you get where you want to be and put you in touch with others who can also help.

Awards and competitions

Other skills-based activities taken on your own initiative are always worth considering – they look good on the CV and impress future employers. These things will put you above the rest and show your potential employer that you are committed to your personal development and an extraordinary candidate. I suggest going in for awards and competitions, my daughter's Young Birder of the Year for example. I won an Earthwatch Millennium Fellowship a few years ago, a travel bursary at University, and recently a National Training Award.

Something like Duke of Edinburgh's award is also a good qualification as you can demonstrate a broad range of skills as well as determination and commitment. Events like running marathons, long distance walks, car rallies (one of my trainees did the Paris-Dakar rally in a car worth £100), foreign travel (backpacking or an expedition), organising events and fundraising all show that you are a dynamic individual and that you have some interesting experiences to talk about. These experiences also give you the opportunity to try or learn new things and meet other like-minded people.

In the next chapter I will show you how you can demonstrate your skills. With relevant, quality volunteer experience, you have something tangible to show prospective employers.

∞∞∞∞

Chapter 6

Demonstrating your skills

It is crucial that you can find the right way to demonstrate your skills and experience to your prospective employer when they are trying to decide whether you are going to be one of the people they want to interview. It is not enough simply to *have* the skills and experience; you need to ensure that you effectively *demonstrate* quantifiable competence in these key qualities to an employer both through your CV and at interview.

The key approach to success in marketing your skills is to put them in the context of the role the employer is hiring for – e.g. even if you have great plant ID skills, you will massively benefit from demonstrating that you have used these skills to carry out botanical surveys. They will want to actually assess your application of skills by having an in-depth conversation with you about surveys you have carried out. They will almost certainly also want you to identify species such as plants or trees during the interview, so be prepared!

Ecological skills

Providing a concrete demonstration of your ecological skills is one of the hardest things to get across on your CV and at interview. This is where you need to call on your hard-earned voluntary experience to show how you've applied your skills in a real-world context.

I will give you an example. When I first started looking into consultancy I found it frustrating that people did not believe that I had strong enough botanical ID skills. I had learnt plants from my mother and my grandfather; I had always had a strong interest in them and learnt new species all the time. But I had no certificates, no-one had worked with me doing botanical surveys and, apart from offering to identify plants at interview, I had no way of demonstrating this skill that I was carrying around in my head. I decided that I needed to find a way to demonstrate my skills credibly so I designed my own volunteering. I approached my local wildlife trust and offered to do botanical surveys on some of their reserves. They welcomed this with open arms and surprisingly did not ask what experience I had - I was willing to work for nothing and that was enough. I surveyed one or two reserves, provided a habitat and species list and presented it to the reserve's officers. I then approached the records centre at the trust and offered to carry out botanical surveys for County Wildlife Site Assessments which I found out about by being involved in the trust in the first place – a good example of being in the right place at the right time, of being involved. The records centre knew that I had done some ID work for the reserves people but they still went out with me for the first day. They ensured I was able to come up with a species list that was as good as theirs, to demonstrate that I could map habitats and that I could carry out an assessment to a satisfactory level. I was offered 5 site assessments at first... then another 24. After nearly 30 botanical surveys I had something pretty concrete to show that I could do botanical surveys - and I got paid for them all! A future employer could see on my CV that I had experience, a certain level of skill and the number of surveys showed that the records centre was happy with my work or they would not have given me that many to do. Demonstrating experience like this is crucial, but don't be tempted to exaggerate, employers can and do check up on an applicant's experience. If you put the time in to create quality volunteer experience that demonstrates your ID skills you won't need to exaggerate and will be happy to list the survey team you worked with as a reference.

Working towards licences demonstrates experience

Another good way to demonstrate skills is to get involved with groups such as your local bat group, amphibian and reptile (herp or herpetology) group or mammal group and start working towards protected species licences. If you start your career with, say, a great crested newt or dormouse licence then that will put you way ahead of the other applicants. If you have a bat licence, which takes much longer and is more difficult to get, then you will be top of the list and will almost certainly get a job straight away! A bat licence is acquired almost exclusively through your own efforts in your own time so you might as well make a start right now. As well as showing the extent of your practical skills, framing your experience in terms of progress towards a licence shows a prospective employer two very desirable qualities: 1) the level of self-awareness you show in reporting how far your skills and experience have developed towards a professional standard; and 2) an awareness of the fact that much of what consultants are employed to work on is directly linked to wildlife legislation.

Getting involved with local groups shows your interest, gets you into the local wildlife scene, gets you known, enables you to learn new skills and gives an opportunity to apply your existing skills. By immersing yourself in these local groups you will mix with a range of abilities and will almost certainly get the chance to work with some amazingly knowledgeable and talented people. Such experts are probably involved with the group because they want to give something back and help others develop the same passion about the subject that they have. What a great way to mix with experts, ones that are willing to share their knowledge. Do get involved, do turn up regularly, do ask questions and show a genuine interest in the subject and in other members. Groups may be suspicious of newcomers if they feel that the new member is only interested in using the group for their own gain and will then leave when they have what they want. If you are going to join a group, keep going along to meetings once you're more experienced and be willing to contribute to the skills and knowledge of newer members.

Getting ID skills accredited

Another way to demonstrate the strength of your ID skills is to gain a certificate of accreditation. This should involve an element of assessment. For example, the University of Birmingham in partnership with the FSC (Field Studies Council) and BSBI (Botanical Society of the British Isles), run a programme of part-time residential courses on ID and biological recording skills. You can tailor the course to fill in the gaps in your ID knowledge or get you started if you have no ID skills. If you're already familiar with plant ID skills you could study grasses, lichens or bryophytes to build on existing knowledge. The scheme provides a systematic approach to species identification taught by experts in their field. You will gain a Certificate in Biological Recording and Species Identification – a tangible series of ID courses with an excellent reputation for quality. You can also just complete ID courses with the FSC - these have a great reputation but don't usually involve assessment. As I mentioned in Chapter 4 there are other accredited courses such as The Natural History Museum's Identification Qualifications in various taxanomic groups University of Leicester offers a similar modular course to University of Birmingham's leading to a Certificate of Higher Education in Plant Studies and Field Botany.

Ensure that if you do a species ID course to gain accreditation for your ID skills that it delivers a recognised qualification based on assessment. That way you have something tangible to show to a prospective employer.

Demonstrating character and people skills

Your interests outside work often show that you have certain skills or character attributes – if you went backpacking for a year, for instance, that shows organisational abilities, confidence, ability to manage money on a budget, resourcefulness and a positive attitude. It also shows that you have learnt some new skills – getting yourself around a foreign country where you may not speak the language (i.e. people skills), how to handle awkward or challenging situations, finding out how to get from A to B, keeping yourself fed and healthy (or even alive), appreciating what you saw, who you met and what you learnt. And of course self confidence – an essential attribute for you and your future employer. You will have grown in many ways by undertaking an experience like this.

Your other interests may demonstrate a range of other attributes or skills such as tenacity, people skills, fitness, winner's mindset, positive attitude, bravery, giving and sharing or specialist knowledge. For this reason, as long as you are being truthful, your personal interests and achievements will say a lot about you, so always mention them in your CV and if possible your interview.

Previous work experience will almost certainly have given you some useful skills. Although I do not encourage you to work in a pub or fast food chain if you can possibly get work in something more relevant to ecology, you can still demonstrate that you have developed a range of useful skills such as people skills, dealing with customers, sales, administration, team-working and self discipline. So, although you may not feel like you are learning much if you are pulling pints or flipping burgers, you will be learning many of the skills I have just listed - all useful stuff for a future employer.

Remember, when you are filling in your CV, to put down the *skills* you have gained from each of your personal achievements and work experiences. It is not always obvious what skills you may have gained from these at first, so give it some thought, maybe ask a close friend or family member to help. It is amazing how interesting you start to sound.

At first it may be hard to imagine how my nursing skills might have helped me as an ecologist – but they did! Working as a nurse I had to build an almost instant rapport with all people from small children, to people of my own age, the seriously ill and the elderly. I have therefore spent a lot of time developing my people skills. I learnt to handle senior staff, junior staff, my day-to-day workload, the paperwork and my own learning all whilst working unsocial hours, being on my feet all day and working in a mentally and physically exhausting job. I had to be cheerful, positive, on the ball, professional and confident at all times, no matter what happened around me or in my personal life. As a professional I was expected to organise my own work and deliver it without anyone telling me what to do and to work unsupervised with seriously ill patients. Wow! What great skills these turned out to be for being an ecologist, running a business, training people and living a fulfilled life.

IT skills

Virtually every business now has some element of IT. Whether it is for the letters or reports you write, your employer's website or blog, an accounts package, spreadsheets for recording or analysing data, a species database or your day-to-day emails – you can't ignore the IT skills which are essential in the 21st century. Demonstrating your IT skills can be hard but usually, if you have recently completed a degree, you will have a good grasp of the basic software packages such as word processors, spreadsheets, email and searching/browsing online. In your communications with your prospective employer they will be looking at the IT skills you have – can you attach your CV to your email? Have you formatted your CV showing that you can use some of the more advanced features of the word processing package? Is it in a standard file type (never send any thing in odd file types like .rtf – always use Word .doc or .pdf as these can be opened easily by almost anyone, anywhere.

In years past people always said that covering letters should be written by hand, but I am not so sure this is the case anymore; after all when do we actually write by hand these days? Employers like to see a neat letter that is well-written, attractively laid out and interesting to read. Once employed you will be writing reports using a word processor as part of your job and your ability to use standard software to do this is a very basic requirement that will be expected in all consultancy jobs.

You may also be at an advantage if you can demonstrate some other more technical computing skills. Consultancies frequently maintain their own websites, blogs and newsletters – have you produced something like this yourself? Of course, if the company you're applying to does produce such marketing material make sure you have read some of it before you get to your interview. Experience of statistics and data analysis will always be considered an advantage, especially if you have used any of the specialist software in use by ecologists. Geographical Information System (GIS), although not used in many types of projects, is sought after at larger consultancies. Often an employer will be interested in adding a new skill to the mix.

Although typing is not really an IT skill, I would urge you to learn to touch type. One of my colleagues is fast but only uses two fingers to type and has to look at the keyboard to do so. Another of my colleagues and I can touch type and so we can type fast, and without looking at the keyboard. This means we can copy text easily, or just type whilst checking what we type is coming out correctly. You can either learn in an evening class or by buying a computer package or book and working through it until it is a habit. I advised my daughter and my son to do this, my daughter took up the challenge and learnt in her school holidays, just a half an hour a day for a few weeks and she was up to speed. It has revolutionised her written work. My son, who was not really interested and could not see the benefits, still types, slowly, with two fingers. It is incredible how much easier it makes life when writing reports if your typing just flows straight out of your head and onto the screen.

Developing confidence

With added knowledge and skills you will, over time, increase your confidence. You will know enough about ecological topics to confidently enter into professional conversations about them. But at what point should you feel you have learned enough to go for ecological jobs?

It depends on what job you are going for, your level of confidence in your abilities and your general level of self-confidence. You have probably noticed that people with a lot of self-confidence seem to get on in their careers much quicker. This is partly because they are willing to take risks, are willing to risk failing and will offer to do something that might take them out of their comfort zone. Some people find this extremely difficult whilst others do it all the time. I have to say that those that fall into the first category and have a willingness to push their comfort zone on an almost daily basis will lead to them having an enormous capacity for growth and a conviction that they can tackle *anything* they put their minds to. Moving outside your comfort zone expands your capacity to tackle new things - it becomes easier if you develop the habit of pushing yourself, accepting that you might fail and not retreating back to your comfort zone if something gets difficult or you fail to achieve what you set out to do.

On the subject of failure: it is rare that you actually fully fail at something. It is more usual that you achieve part of what you set out to do but not all of it. Isn't it better to have achieved part of a new goal, to have pushed your comfort zone, learnt something new and achieved *something*, than to not have tried at all? Your attitude to failure could be a major stumbling block in your career if you don't recognise it as a potential limitation and work on *yourself* to ensure that you don't excessively focus on what could go wrong.

When you start your new career you will not know everything, so don't expect to know it all when you are starting out. In my experience the more I know the more I realise I don't know! We are always learning about new subjects and about ourselves; this is part of making your life and your career interesting. Accept that there are people who will always know more than you – get to know the experts in your field professionally and personally. Having a network of colleagues and friends who you can ask for help is critical to succeeding in a career as diverse as ecology – you don't have to know everything!

∞∞∞∞

Chapter 7

∞∞∞∞

Your Curriculum Vitae (CV)

and covering letter

Your CV might be the first contact you have with your prospective employer. If they are anything like Acorn Ecology then they receive CVs all the time, or if they advertise an ecologist post they will probably receive hundreds of CVs. How do they decide whom to interview? With a vast number of applicants and only one job it is very difficult to decide between candidates' CVs. So how do you make yours stand out? What can you put in your CV that will make them want to know more or to interview you?

The first thing to note is that you don't really know what they are looking for so you might as well just be honest and not exaggerate anything you have achieved. Having said that however, don't be afraid to highlight your achievements –

no one else will (apart from your referees maybe). They will know if you have the right balance of skills they are looking for when they read your CV. These might not be purely ecological skills - for instance, when I was looking for a trainee ecologist I also wanted the applicant to have strong admin skills so they could help with day-to-day business tasks. I didn't interview anyone who had never worked in an office, even though they may have had some of these more generic skills but left them off their CVs.

Some people try gimmicks to attempt to stand out from the crowd; things like coloured paper/colourful design, a photo, or a link to a website where there is a video of you speaking. Generally the standard CV will be read through for content and the gimmick won't make much difference. The substance i.e. qualifications, qualities and attitude are what they are looking for. Remember to keep your CV to no more than 2 pages – 1 page is not enough and 3 is way too many.

Make sure your CV is printed on good quality paper and the format is balanced – avoid crowding text or too much white space. Make sure your spelling is correct – use the spell checker but don't rely on it as they don't identify all mistakes (for instance the word may be correctly spelt but in the wrong context e.g. their and there). Use appropriate language which communicates the information clearly and concisely.

What a CV should contain

If you are not sure how to organise your CV I give an example of suitable headings in Appendix 2 and in the following paragraphs I'll walk you through the way to write your CV section by section. Remember, just like any other CV, you need to tailor the content to the job you are applying for. Your CV needs to be carefully focused on your next job as a professional ecologist.

Header information

This needs to contain basic information like your name, address including post code, telephone numbers you can be reached on, email address, date of birth (yes you can include this if you want – it is against the law to discriminate against you on age so you should not worry about giving this information) and an indication that you have a full driving licence. If you don't have a driving licence I would strongly advise you learn how to drive – it will be very difficult to work as an ecological consultant without the ability to drive to field sites. Most consultants would be very concerned that a non-driving applicant (for a job involving field work in remote locations) would need constant ferrying around by colleagues.

Summary of skills or personal statement

It has become fairly standard to put either a short bulleted list, or single paragraph summary at the top of a CV. Keep it brief – what 4-5 main points do you want to get across about you? For example 'a passionate botanist with extensive experience of UK flora', 'confident public speaker and presenter', 'life-long birder and member of the BTO/RSPB' etc. Try to avoid generic statements about team-working etc.

Education and qualifications (GCSEs, A-levels, degree, post-graduate degree)

Make sure that this section is not a boring list of GCSEs, A-levels, and a degree in a list. When mentioning your qualifications it is usually sufficient to say how many GCSEs you got rather than listing them all, the grade range and if it included Maths, English and Sciences. Do list your A-levels though, as this begins to show your subject interests, your grades show your calibre although you can omit grades if you are not proud of them... as long as you passed. Always put in the grade you got for your degree (e.g. 1st, 2:1, 2:2), employers are interested in that and if you don't put it in they will only ask or assume it wasn't great.

Work experience

If your work experience has been a series of burger bar, pub or shop jobs then don't just say where you worked and for how long but focus on the *skills* you gained. For example saying you worked for the local pub doesn't really tell anyone anything. Maybe you learnt to handle people, to work in a team, how to handle money, how a business works, how to manage staff or do staff rotas, maybe you organised quiz nights or other events, or stood in for the manager. Skills are what we are looking for. People and team skills are probably the most important to demonstrate so make sure you mention this if you had a customer-facing job. The length of time you stayed in a job is also telling. It is good if you can show some staying power. If you have worked in something unusual, or abroad or something you initiated yourself then talk about that, it shows you are a bit different, maybe more motivated or willing to take the initiative.

As you get older your CV could be rather long so you may have to lump your working life up a bit. My nursing experience ended up as the 'from' and 'to' dates, a brief outline of the jobs as a whole and my main duties and responsibilities. Ten years were summed up in a line or two. Always make sure that all your work experience is listed with no chronological gaps, you may be asked what you did in those gaps so you might as well explain them in your CV, even if they were a gap year or volunteering, you were still gaining skills. Your work experience and education should therefore span the dates from school to the present with no gaps. If you have an overlap with education and work, then explain that the work was part-time, if it was. I actually had a year of full time work whilst doing my part-time MSc so there was an overlap there.

Ecological training and experience

This is one of the most interesting sections for prospective employers – it's here where you can list your specific ecology training, voluntary experience and wildlife groups you've been active in. You should organise this section exactly the same way you did the education and work experience sections, with dates, places and names of courses/groups/organisations. If you have been a member of e.g. a herp group, then list the date you joined to the present to indicate that you're still a member (you are probably still a member). You should then list what you've learned as a member of the group, the activities and/or surveys you've undertaken and possibly the names of any well-known ecologists you've worked with if that seem relevant.

Achievements

In your Achievements section list any other notable things you have done – grants or awards, organising a group event or trip. If you have a blog, or an article that has been published online (if it is in some way related to ecology) list it here and link to it directly. Having some written work published is a major achievement and a huge plus point for employers – it demonstrates that you can write and that you've been tenacious enough to find someone to publish your work. You may be struggling to think of something for this section and if you really cannot think of anything then leave it out.

Hobbies and interests

Include your interests on your CV. There is a school of thought that suggests CVs should be all business; that only professional experience counts. That might be the case in the financial services industry, or the legal profession, but an ecology employer wants to know about YOU, they won't want to just read through a dry account of your education and practical skills. For my first trainee ecologist is was this section that got her the job. Everyone who applied had virtually the same education or work experience but the fact that she was into mountain climbing, kayaking and art clinched it for me. It showed she was fit, adventurous, and creative. Excellent qualities in anyone, especially someone you're hiring. She had also been to South Africa to help AIDS orphans with her church group which showed her compassionate nature and that she had experienced life outside our cosy existence here in this country. Another trainee I took on listed reading, arts & crafts and cooking. She was lovely but not fit enough for the really tough aspects of fieldwork. One trainee did not put in her interests, when I asked her at interview she said she tamed horses and rode a motorbike! She suddenly sounded much more interesting. Your relevant experience and personal interests should show commitment, enthusiasm and passion.

It is your personality, your interest in the subject, and often your fitness that is what we are really looking for. Most people applying for these jobs have the same qualifications and many of the young ones only have part-time jobs as their work experience. For the more mature applicant you still need to demonstrate skills – what skills did you pick up when you worked as a quantity surveyor for 20 years? How did you progress in your career? What did you do in your spare time? What did you do to get you to this point? Hopefully you have studied a relevant degree or relevant courses, volunteered for the local wildlife trust or got involved in something relevant.

References

If you have just graduated then it's typical to have your university tutor as one reference and your most recent employer from part-time work as the other. References from people who have worked with you are invaluable. Are there any ecologists you've worked with on surveys etc (while gaining you valuable experience)? If you have interacted with someone then ask if they would be willing to act as your referee for a job application. An ecologist, especially if they can vouch for your ID or fieldwork skills, will provide much more credibility to an employer than a person you did part-time bar work for 2 years ago.

An example of a poor CV made good

A close friend of mine who had recently graduated was applying for her first job in ecology and asked if I could look over her CV for her. I asked what the job was, what she would be expected to do in that job, and then I read through her CV. I became rather despondent as I read it - it all sounded rather pedestrian, nothing was standing out for me. A string of GCSEs, 3 A-levels, a couple of part-time jobs, a degree, some interesting hobbies but this was not the person that I knew. Perhaps she was having an off day when she wrote this. I re-wrote parts of it and when I sent it back to her she couldn't believe it. 'I sound amazing!' she said. 'You ARE amazing!' I replied.

I asked her why she had forgotten to mention that she was fluent in Spanish, French and Portuguese, had backpacked around Thailand on her own, had organised an 8-week trip around Brazil had had done lots of relevant voluntary work. She had also achieved the rank of Petty Officer in the Naval Cadets at School thus demonstrating strong leadership. She had won two awards and she had omitted to tell the reader about the skills she had acquired whilst doing her seemingly menial part-time jobs.

Think about yourself, your skills and achievements. We have all done extraordinary things in our lives and we have opportunities that generations before us could never have even dreamt of. If you are a bit stuck about what to write ask some long-term friends or your family to remind you of what you have done and how amazing you are. Where possible put over to the reader of your CV your enthusiasm and commitment.

Your covering letter - tell your wildlife story

The covering letter is often treated as something of an after-thought when applying for a job. However, this is a huge mistake as it's one of the best chances you have to get across some of your personality to a prospective employer. Being far less structured than the CV, the covering letter gives you a fair amount of canvas on which to sketch out your story.

I don't really like it when I get 'Dear Sir' letters when it is pretty easy to find out from our web site that the senior ecologist/managing director is a woman (and that I have a name!). Where possible address your covering letter to a specific person. If you cannot find out who that might be then ring up the company and ask whom the best person to send it to is. The covering letter should introduce yourself, where you are in your career, what you are looking for, why you might want to work for that company as well as what you can offer them. I love it when people send in a letter that shows they have found out about what we do – make sure the covering letter is tailored to the company you're sending it to. Each consultancy is truly unique and exploratory research might also help you decide which companies you don't want to send your CV to.

Make sure your covering letter is well written, appropriately punctuated with capital letters used correctly. Don't ignore the spelling and grammar checker on your word processor and make sure you have the spelling set to UK English not US English. Bad CV and covering letter presentation sends a very bad signal to a prospective employer. If an applicant is not careful about spelling and layout for their own job application then they are likely to take even less care with their work and, after all, a major part of their work could well be preparing reports, if not at first, then certainly later on in their career.

If you have no relevant experience, either paid or voluntary, then you might find it difficult to know what to say in your covering letter. You should, no matter how experienced or inexperienced you are, have a way of demonstrating your passion for wildlife. That's what I call your 'wildlife story'. What's yours? Try to give the reader a sense of your passion for wildlife and your understanding of how being a professional ecologist fits with that. Despite the rigorous professional nature of a modern ecological consultant's work; the legal aspects, the report writing and business side, the majority of ecologists enter the industry because of a passionate interest in flora and fauna.

Sending off your CV in response to an advertised position

There are two main ways you can use your CV. One is to send it as an application for a specific job and the other is to send it to any companies or organisations that you are interested in working for.

Before you send off your CV for a specific job, always review it to ensure it is relevant to the job you are applying for - make it as specific as possible to that particular job. Sometimes you might want to emphasise something a bit more, any experiences you have had that are the most relevant – particularly if they have requested certain skills or experience that you need to demonstrate. Make sure the CV arrives on time. Employers are unlikely to consider any CV that arrives after the advertised deadline.

Sending a speculative CV

Many people only send CVs to organisations when that company is advertising a job and never send out speculative CVs off their own back. I have never advertised a job as I get enough speculative CVs to make a good choice – many consultancies find that this is the case. If you have never sent a speculative CV there are likely to be many opportunities out there that you would never be aware of.

Sending your CV to an organisation speculatively should be something that you do as a matter of course, particularly if you want to live in a certain area of the country. It also shows initiative and motivation. Hopefully the employer will reply to your letter and CV saying they have kept your CV on file, if not, it is worth checking if they actually received it. I have kept good CVs for many months - one of my star employees submitted his CV around 7 months before I got in touch. He was not expecting a call but was delighted when he did, and he got the job! His CV impressed me so much I kept hold of it. Sometimes I also contact people who have sent in their CVs to invite them on a course or to attend an event or help out with surveys if they are local.

Keep in mind that ecology is highly seasonal. Most people are looking for recruits between April and end of September, so send your CV early in the season – March is ideal - for the best chance of getting something. If you send it in October you are unlikely to hear anything for some time.

In the next chapter I will cover what to do and to expect at your interview. If you are invited to interview you know that you and the way you presented yourself in your CV are just what they might be looking for – you'll want to reinforce that message when you meet them in person.

∞∞∞∞

Chapter 8

Successful interview technique

If you are asked to attend an interview then you know that your CV and covering letter have done a good job, that you have the necessary qualifications and probably the right experience. The company you've applied to likes the look of you on paper and are now keen to meet you to find out what you are like in real life. They may want to test you on some of the things you say you can do. The other main aspect of interviewing candidates is to choose between people who all look reasonably good on their CV. Often, at this final stage, the front-runner from the CVs can end up looking a little dull compared to someone at interview with more enthusiasm and interest who did less well in the CV.

Before the interview

If the interviewer has asked you to prepare a test piece of research or written work prior to the interview then make sure you do it well and submit it on time. This is fairly common practice. If you have been asked to confirm attendance at the interview, then do that too.

To prepare for the interview I suggest you look up the company or organisation - they pretty much all have web sites. Find out who the senior staff are, search for articles or books they may have written, or look at products they make or fields they specialise in. It may be that they are nationally known for something and you didn't realise. Maybe they are the Chairman of an important charity or steering group. Find out if it is a family run business or a large multinational, the difference between being a key worker in a small company and being a tiny part of a large machine should be something you should also consider. What sort of organisation do you think you will fit in with best, or indeed would prefer to work in?

Being able to find out all of this online makes this research process so much easier than it was even 5 years ago and for that reason it is expected that you would have put in some effort to find out more. By getting to know the company or organisation as much as possible before the interview you will be able to formulate some questions or areas you're keen to cover in the interview. If there is a really burning question about the role you could always email or ring them up to find out more before the interview, they probably won't mind.

Make sure you read the job description, and person specification if there is one, before the interview so that you can ask any questions and also to double check you have all the qualities and skills they are looking for. When I went for an interview with a charity that works with barn owls, part of my job would be to put up owl nest boxes. I was really confident that I could do all the other things in the job description but I had never carried a barn owl nestbox up a ladder. I did not want to flunk that bit as I was expected to attend an all-day interview, so I phoned the boss and explained the situation. He said I was a very strong candidate and was willing to come to my house, put up a ladder and let me have a go! So this I did. I found it surprisingly easy, even though I am only small. I was very grateful that he gave up his time. So I attended the interview and got the job, which I enjoyed immensely.

By checking the company out and the job description it is always possible that you wouldn't want the job anyway. I found it better to go for jobs I really wanted rather than wasting my time, and theirs, for a job I either didn't want, was not suited to or not going to enjoy. The pay and hours are also a consideration. You might think the pay sounds good but are they going to give you overtime as days in lieu or will you be expected to work all hours, especially in the summer, for the same money? This is an unfair but extremely common practice, one that you might be prepared to put up with to get a foot in the door but not to set a precedent for the rest of your career. Ask around to see if the company has a reputation of being good to its staff, this might not be something that the company itself would admit to, especially at interview. You may be able to find out the staff turnover which is a good guide to job satisfaction.

Preparing yourself

Even to the seasoned professional who may have endured dozens or even hundreds of interviews it is still a daunting process. It is easy to let your nerves get the better of you and for this to affect your performance. Being armed with the information you have gathered about the company, the job, and probably the interview process, should make you feel a bit more confident.

I have been on several interviews (which actually tended to be for charities - really badly paid but really interesting), that lasted all day or even more. So be prepared for an assortment of activities, not just a straight interview. Once I had an all day interview where I had to do map reading, give a presentation, do a group exercise with other candidates, also an ID skills exercise, a woodwork exercise, put up a ladder, go up a ladder, meet the staff and have interviews with three people. I went on another one for a job as an education officer at a zoo where I had an interview, had to give a presentation on something I prepared in advance, then another presentation on a topic based on a prop that I had to choose, also a group-work exercise and taking a group of children for a short session. They even made me handle some mystery live animals (I did not know what they were until I opened the lid – they were hissing cockroaches) and then had to speak about them for 3 minutes! Then I had lunch with my interviewers and fellow candidates. All pretty traumatic and exhausting and I was left wondering what on earth they were going to ask me to do next (and I didn't even get the job!).

Wanting the job is a good start but wanting it too badly can also hamper your performance. I was once turned down for a job and when I asked for feedback was told that 'you didn't seem to want it badly enough'. I have to confess I was a little lukewarm about it and I know the woman who got the job and she is full of enthusiasm and confidence, so I can see why they chose her. My first trainee wanted the job so badly that she was visibly shaking at interview and could hardly speak. I decided that I liked her though and invited her to spend the day with me to go out and do some survey work and some work in the office to test her in a less stressful setting. If I had been interviewing 6 or 8 people I probably would not have bothered. On this second interview, where it was a less formal situation, she was fine and impressed me with her knowledge and skills.

My interviews for trainees always include a chat, an ID exercise (usually plants) and then maybe an admin or IT task. If I have time I also like to spend the day with the candidate to see if we get on.

To help you with confidence, if that is a problem for you at interviews, there are some clever psychological tricks you can use. The first one is to imagine the interview the night before when you are lying in bed peacefully and relaxed. See yourself cheerful and confident, shaking hands in a friendly manner with your interviewer, chatting to them with ease. Play the interview through in your mind, feel what it is like to be there and feeling confident. Next, see yourself with the letter accepting you for the job and how elated you feel. See yourself turning up to the first day at work and how you will conduct yourself. Top athletes use this technique to visualise the whole race before they actually run or it, including visualising holding up the trophy at the end to thunderous applause. It might sound a little crazy but this visualisation process is known to work – it's called neuro-linguistic programming (NLP). It's very popular with high-achievers and well worth reading up on.

Another technique is to work on how you are feeling on the day. If you are starting to feel nervous, you can't eat or you are starting to find it difficult to smile then try this technique. First, stand up straight and take a couple of deep breaths in front of a mirror. Smile at yourself and tell yourself that you are feeling great and you feel happy and confident. Keep smiling; try a little wiggle to loosen up your body, maybe a quick dance to your favourite music. Keep telling yourself you are happy, confident and enthusiastic. Smile and keep standing up straight, walk around a bit as if you are an extremely confident person, the one you visualised the night before. Every time that little voice inside you starts to put some doubts in your mind just push it away and ignore it. Concentrate on your feeling of confidence, enthusiasm, happiness and well-being. You see your conscious mind can be fooled into thinking whatever you wish, even if it is not true, and your subconscious mind listens to your conscious mind. It can only process now, it can only process one thing at a time, and it can only process the feelings coming from the conscious mind. The subconscious mind is the one bringing up nags that might deter you but you must persist. In only a few minutes (5-10 minutes), with constant vigilance to the little nags (which you recognise and instantly discard), you will start to feel really good. So walk confidently, and keep smiling!

Enthusiasm is infectious and is an expression of a positive mental attitude and well worth nurturing. Practice enthusiasm in your normal conversations; maybe try reading aloud enthusiastically for 10 minutes a day in front of the mirror.

At the interview

Keep smiling - be confident and enthusiastic; it is most important that you keep this up right the way through the interview. Shake hands firmly but not too hard or too limply. Some people read a lot into a handshake! Show interest in the other person, your enthusiasm will be infectious, eagerly answer questions and have some to ask too.

Most of the people interviewing you are not likely to be professional personnel managers but just people trying to run a company or a department. However, if you are being interviewed for a very large organisation you may have a professional personnel person present. They may have a more formal way of speaking or come up with seemingly random questions that seen unrelated to the actual job. I have heard of people being asked 'If you were a biscuit what sort would you be, and why?' Personally I find these questions ridiculous and difficult to answer, there is no right or wrong answer but often they throw you so much that you may get flustered. Just go with it and do your best, maybe they are looking for someone who can think on their feet.

Larger organisations may include some aptitude tests or similar, again don't try to answer in a way that you think they want you to, you might as well be honest. If you are not right for them then there is not much you can do about it and it is better to let them be the judge, as they know the job and the organisation better than you.

For most jobs you will go for it is likely that your future boss or senior colleagues will interview you. Remember that they are interested in you already or they would not have invited you there. The interview may just simply be a process of deciding which *person* they like the most or think would most suit the job and the team. There is not much you can do about this apart from being yourself. If you are confident, presentable, friendly, bright and enthusiastic that will go a long way towards getting you that job.

Your appearance

On the day make sure you turn up on time or 5 minutes early, if you get unexpectedly held up then call them to let them know where you are or what has happened (e.g. heavy traffic or flood or an accident on the motorway). Wear smart clothes, but not super smart (a suit and tie is probably a bit over the top for most ecology roles). I have had people turn up for interview in scruffy clothes, with blue hair or dreadlocks, or with numerous piercings around their face or with skimpy revealing dresses. None of which impressed me. In fact these interviews were rather short, as even if they were the nicest person in the world or had great experience, I would not have employed them. These people will be representing my company, including talking with clients (our livelihood), and they gave me the wrong impression, so I was concerned that they would also give my clients the wrong impression. Sorry if this sounds old fashioned but this will be the case for most employers.

Interviewers are often selecting on whom they want to represent their organisation or who would work well in the existing team, you have already demonstrated your qualifications or experience. Your fellow interviewees will probably have the same or similar to offer, and particularly for the more junior jobs where the company expects to train you up in their particular way of doing things.

You may be shown around the office and get a chance to meet the other staff. Be friendly! They may all be asked what they thought of you after you have gone.

After the interview

Thank the interviewers for seeing you; shake their hands firmly and with a pleasant smile on your face. Ask them when they expect to make a decision. When you get back from the interview you could send an email thanking them for the interview and saying that you enjoyed meeting them and discussing the role. Most people don't do this and I think it's a massive lost opportunity.

If you get a call that day that it's usually a good sign, unless they have a reason to wait to make the decision. Usually they will offer the job to the one they preferred on the same day, and if that person turns them down then they can fall back on the next one without it seeming like a long time since the interview. This also helps employers disguise the fact that a person may have been a second choice. They may also have interviews spread over several days. If you have not heard for a couple of days after the day they said they would let you know then phone up and speak to the interviewer. This can be difficult as you have probably not got the job, but be friendly and courteous. Ask for feedback so that you can learn for next time.

If you do get offered the job then congratulations! Hopefully you will be delighted so make sure the interviewer knows that. They will be looking forward to working with you as well. Ask what happens next. You may need to give notice at your existing job depending on what sort of work you are doing at the time and if you are under contract. Usually you will agree a start date and time and then they will send you a letter confirming your appointment. Your offer will usually be subject to them checking your references – they will call or email your referees. You may have to provide some documents such as proof of qualifications/licences, identification and National Insurance number - you may even have to have a medical. They will inform you of what to bring and what to expect for your first few days.

Prepare yourself for your new career by reading up about anything you may have picked up from the interview – or ask them if there is anything you should read before you start. If the company specialises in anything you're not familiar with then it's a good idea to try to learn something about it before you start. Welcome to your new job!

∞∞∞∞

Chapter 9

∞∞∞∞

First Job and beyond

When you turn up for work on the first day of your new career in professional ecology you should have a pretty good idea of what your job will involve from the job description and the questions you asked about the job at interview. However, your first few days and weeks are still a settling-in process and you might not do anything particularly interesting for the first few days. During this time you will start to get a feel for the company culture, standard procedures, who does what and who to ask for help on different topics. Before starting any real work in most offices there are policies and procedures that may have to be read, health & safety training briefings, employment contracts to sign and colleagues to be introduced to. Once your first few days are over you will get the opportunity to go out and do what you are paid for – surveys, site assessments, mapping, translocations as well as the office stuff like writing reports, logging your hours, preparing invoices, administration, writing letters, answering emails, doing quotes. How much you get involved in the office activities generally depends on the size of the consultancy and how much survey work they have. Larger consultancies often have a team of admin staff dealing with the non-ecology side of the business but in smaller consultancies you will almost certainly have to muck in with everything, including the office work and helping to run the business.

As a entry-level ecologist you will spend a lot of time out in the field (and that is what most of us want to do), working on the more straightforward but time-consuming jobs such as reptile surveys or translocations – this allows the consultancy to free up senior staff to work on more technical projects.

Set the tone

Start your new job as you mean to go on. You will not know anyone, except perhaps the person that interviewed you, so you need to make a good impression. Be yourself, after all, your employer liked what they saw in your interview but, remember, that in all new jobs anywhere, you are effectively on probation in the first few months. When you start you will have to slot into an existing team of people who already know each other - a positive attitude, a friendly manner and a willingness to do anything requested of you will make a good impression and help you settle in quickly. Above all else, remember that unlike university or volunteer placements, you are now being paid to work – you're a professional in an industry where building a good reputation is essential to developing your career.

Hopefully you will have a meeting with your employer early on where you can discuss your plans for the role, your objectives for the next few months and any worries you may have. In my experience it is a good idea to have a clear idea of what you want out of the job so that you can ensure you have a focus and your employer knows what you want to do. Bear in mind that your employer has not hired you for the purposes of training you to do exclusively what you're interested in. You need to make sure that you align your objectives with the projects and plans of the consultancy you work for rather than expecting them to fit in with you.

Being keen to learn is an incredibly important thing and will make the difference between you getting on in your career quickly or slowly. If you're keen to gain experience in a new skill then pick up what theoretical knowledge you can, usually through background reading, before trying to get someone to pay for your training – it sends a clear signal that you're sufficiently interested and keen to master the technique and that they won't be wasting time trying to get you trained up.

Getting the most out of your first year

Starting with a focus will help you and your employer make sure that you get the most out of your first year. This is the most intense time of learning and confidence-building for you in your new career. Get involved with anything and everything, ask the senior ecologists questions and discuss projects. If you are spending a long time in the car with a colleague then spend the time to increase your ecological knowledge, it is an excellent learning opportunity that could be easily wasted if you just sit and listen to the radio. Observe how senior colleagues work, how they act around clients, get a feel for the professional way to handle yourself in different situations.

Positive mental attitude

Don't miss any opportunity. If you are not invited to a particular survey that sounds really interesting then offer to go along in your spare time if necessary. You can only grow if you stretch yourself - if you stay in your comfort zone you will be limiting yourself, closing off your mind and restricting your career. Learn to push your professional boundaries within an environment of support from senior colleagues – they can guide you in the right direction.

A positive mental attitude will make your work more enjoyable - being positive is simply a decision you can make and then stick to. Your mental attitude is completely in your control - make sure you review your mental attitude and your approach to your work – it is easy to fall into old habits if you are used to being negative or finding fault. With a positive attitude you will enjoy your work and learn much more, get promotions earlier, work better with colleagues and feel more confident. A positive attitude is infectious (as is a negative one) so you can help your team to become more positive too.

Be prepared to work very long hours in the summer and hopefully you will get your time back in the winter. The work in the summer just has to be done and while the hours are long and hard. It should still be enjoyable though and you'll miss it in the winter months. I take my holidays in the winter and cut back on my hours so that I can catch up with my hobbies and life plans. If your employer does not pay overtime then suggest they let you log your hours and then allow you to have them back in the winter. Be careful not to be exploited, this is becoming rather common, especially in some larger consultancies. You are really entitled to work only the hours you are contracted for and you should get back any overtime as time in lieu, or at least a good proportion of it. Otherwise your hourly rate can start to seem rather puny, you get over-tired, your social and family life suffers, and you will start to resent work.

The habit of professionalism

Ecological consultancy is a profession, though you may be only at the beginning of your career, you will soon be a fully-fledged professional, so it's important to behave like one right from the start. In your dealings with colleagues and clients be helpful, courteous and be willing to admit you don't know rather than giving poor advice - you can always refer the matter to your senior colleagues. As you learn and grow, your confidence will increase and you will *feel* more professional. In the meantime accept that you don't know everything and that you are still learning. Being professional is more about your attitude when you start out.

Standards set out by IEEM for conduct, as well as standard survey and mitigation methodologies set out by experts in the field give you a framework to work to. If you do not carry out surveys using correct methodology and to the standard required then your surveys and reports are unlikely to be acceptable to your employer, much less the client. You risk losing professional credibility, surveys may need to be repeated at great cost and projects could face severe delays due to the seasonality of surveys. Make sure you are clear on the standard methodologies either from reading under the guidance of colleagues, from attending courses or by colleagues showing you and explaining them. Make sure you are certain of what you are doing for a survey before you go out to do it. Using standard methodologies ensures that we are all performing our surveys to the highest standards. Having said this you might, after some considerable experience, question some of the methodologies and might even trial and develop new ones that could become widely accepted – but this is for later in your career – perhaps something to look forward to!

Being professional is also remembering that you are working for a client and it is your job to ensure that they are complying with the law with regard to wildlife and that they are provided with the information in the right format to help with their planning application or other requirements. We can cause delays in projects if we miss something and consultancies can be sued if we are negligent in our work.

Continuing Professional Development (CPD)

If you are a member of IEEM you need to log at least 20 hours of CPD per year. Personally I find this very easy to achieve and I wonder if it is a bit low - I regularly log well over 100 hours a year. To ensure your career and your development as a professional is progressing, IEEM set this standard to ensure everyone keeps up to date. This profession is constantly changing and it is important to learn updated survey methods, understand the implementation of new legislation and study the latest research about the conservation and ecology of species you work with.

CPD can include courses, reading, in-house training and conferences. I attend at least two conferences a year - this is a really excellent way to find out about what's new, to hear about research, new initiatives, the latest survey methods, changes in the law and much more. It is also a great opportunity to meet people and start new professional relationships. Between presentations and during coffee breaks you'll be able to browse exhibition stalls to pick up new publications, equipment and information. 14 hours of CPD can be covered in one weekend conference!

Keep a log of your courses, conferences, training and reading so that you can submit your CPD record at the end of the year. It is interesting to look back on your year and see how much you have learnt and grown in 12 months.

Your CPD should also be your opportunity to plan your learning. Why not write a list of the skills you need, information you are lacking, experience you would like to develop in the year ahead? You can then plan your courses, conferences and reading so that they fill in the gaps, that way you do not end up doing CPD that is actually a repeat of previous years. At your annual review/appraisal with your manager you can discuss what aspects of ecology you would like to concentrate on next. It is easier then to decide which courses to go on because unless they align to your plan you won't need to go on them, or at least not just yet. For instance when I was trying to get my bat licence I went on a bat ecology course, a bat surveying course as well as the national bat conference held by Bat Conservation Trust, I also listed the books I needed to read and the experience I had to get and then just worked my plan. Focus on the end result, knowing what you have to do to get there and work steadily towards it.

Start a library

Very early in my career I had already amassed many useful books. You really need to start your ecology library as soon as possible. In the bibliography I have given some suggestions on what your library should contain (see also Bibliography at the end of the book).

The books you need are in the following broad categories:

- *Identification books* – flowering plants, mammals, insects, birds, trees, grasses, amphibians and reptiles.
- *Species ecology books* – I love the Whittet books on various species for a great overview on ecology, field signs and conservation. There are also some real classics such as Ernest Neal's *Badgers* and Hans Kruuk's *Otters*, David MacDonald's *The Encyclopaedia of Mammals* and Harris and Yalden's *Mammals of the British Isles* (you can see I am biased towards mammals!).
- *General ecology books* – books on ecological census techniques, habitat management, conservation biology and habitat creation.
- *Conservation Handbooks* – produced by various specialist organisations – Dormouse, Great Crested Newt, Water Vole, Lesser Horseshoe bat.
- *Manuals* – Herpetofauna Worker's Manual, Bat Worker's Manual and Phase 1 Habitat Survey Field Manual.

You obviously don't need to get these all at once but as you go along, and as your interest in certain subjects increases you can gradually buy them. Some publications are on the internet now so you can download them and keep them on your computer but most are still physical books. Don't just get the books either - *read* them too! I always write down any books that people recommend and get them when it is appropriate to. If there is a library at your consultancy then you should be able to borrow some over the weekends.

Other essentials

Invest in a decent hand lens – you'll need it to train yourself in ID, especially plants, a x8 or x10 is fine. Binoculars are useful for wildlife watching including birds and other elusive things like dragonflies and skittish lizards – 8x40 are good, if they are too magnified then the shake of your hands will be very obvious when you look through them. If you want to get into bats then I recommend you start with a simple heterodyne bat detector like a Magenta or a Batbox whilst you are training, all of a sudden you will find bats that you would otherwise never realise were there.

Make sure you get some good wellies, you'll be spending a lot of time in them. Outdoor clothing including a good set of waterproofs is essential, quick-drying trousers are also very useful. On the subject of wet weather a weather-writer is very useful – this is a clipboard with a plastic hood over it, they come in A4 and A3 sizes. You will gather lots of other equipment over time but for now these items are a good place to start.

Make a plan

Most people do not have much of a plan for their careers. In fact it is said that people spend longer planning their holidays than their lives, longer planning the birth than the childhood, and longer planning the wedding than the marriage!

Where do you want to be in 10 years? For example – running your own consultancy could be your 10 year goal. Then a 5 year goal, to get to that 10 year goal, might be - senior ecologist specialising in bats and badgers. Write it down. Starting with the end in mind makes you far more focused. Once you have a clear picture of your destination your journey is much more rapid and directional, you can then start to plan and achieve the steps to getting there – it is like going on a car journey, knowing where you are going and taking a map. Without a destination in mind you would not even know which way to turn out of the drive! As we have already discussed in the CPD section, you can now plan your training each year to align to your long-term goals.

I have a technique for planning the various aspects of my life and since I have been using it I have achieved 100 times more than before and have become totally focused on achieving those goals. I have given a template for my Life Planner in Appendix 3 and I have prepared an example below.

Start by writing down your 5 year goal and then write down your assumptions and where you are right now. Then start to work out where you will need to be in 4 years, 3 years, 2 years, 1 year, 9 months and 6 months. Finally write down what you need to do in the next 3 months to start getting you to that end goal. Although your 5 year goal may seem massive and difficult to achieve, you simply need to take small steps consistently over time in the right direction to achieve your goal. Set the goal, plan the steps, then work the plan by taking action. I have an example here (read the 5-year goal first, then the assumptions and current situation, then the steps between):

Life Planner	
Career – Ecological Consultancy	
Timescale	**Actions/Aims**
5 years	Senior Ecologist – earning £50k pa, expert in bats, bat licence trainer, nationally recognised bat expert. Expert botanist, ornithologist, herp worker, mammalogist. Expert in badgers with extensive experience in badger work. Teach badger courses to consultants. Competent in all aspects of ecological consultancy including botany, herps, birds, Phase 1. Several staff under me including trainees. Courses x2. Conferences x3.
4 years	Become bat licence trainer. Set up bat course. Set up badger course. Mammal conference, bat conference, bat trainer's training. Botany training in grasses, sedges, mosses, rushes, water plants. Courses x2.
3 years	Promoted to Ecologist. Know most plants, trees, common grasses, ferns. Know most birds. Bird song course. Started doing European Protected Species Licences for bats, dormice and great crested newts. Bat conference, dormouse conference, mammal conference. Now know enough plants to be able to ID most flowering plants and trees and common grasses and ferns. Courses x2.
2 years	Gain bat licence. Done 3 badger licences myself. Know 500 plants, 50 trees, 20 grasses, 10 ferns. Bat conference, herps conference. Badger mitigation course. Bat mitigation course. Bird course, out bird-watching every week in own time with local group. Reading bird books.
1 year	Promoted to Assistant Ecologist. Involved in first badger licence. Gaining bat experience. Great Crested Newt course. Know 250 plants, 30 trees, 5 grasses. Got dormouse and great crested newt licence. Going out on scoping surveys unaccompanied. Writing own reports. Invasive species course, bat conference.
9 months	Learn 100 more plants and 10 more trees. Badger course. Dormouse course. Go out on one badger job and discuss with colleagues. Learn badger ecology and field signs (read Ernest Neal). Great crested newt and dormouse surveys x 6. Read GCN and Dormouse Conservation Handbooks. Read 2x bat books – ID and ecology. Writing reports to include advice.
6 months	Learn 90 more plants (one a day) and 10 more trees. Bat course. Botany course. Go out at least

	once weekly outside work doing bat work. Great crested newt and dormouse surveys x6. Start to read Bat Worker's Manual – conservation, law, health and safety, bat ecology and surveying. Start writing reports. Doing simple surveys on my own.
90 days	List the skills I need. Book training on Phase 1 habitat, and surveying for protected species. Carry out as many surveys as possible for a variety of species – bats, dormlce, badgers, great crested newts, reptiles, Phase 1. Learn 20 trees and 60 plans (two every 3 days). Join local bat group and go on events. Read Phase 1 handbook.
Current	I have done a degree and have just secured my first job as a trainee ecologist but have no experience. I have poor ID skills, especially plants.
Assumption s made	I am intelligent, hard working, committed, and enthusiastic. I have every reason to succeed. I love ecology and I am really interested in wildlife.
Instructions: Write down your aim for where you want to be in 5 years time and write it in the first section under Actions/Aims. Write down your current situation and assumptions made. Then think of the milestones you would expect to be at for each of the timescales given and write them down. What can you do in the next 90 days to start your journey? You need to start NOW! Enjoy the process then TAKE ACTION! Review every 90 days.	

Although this is only an example, if you read it you can get some idea of how having the end in mind helps you to see what you need to do to get there. Each 90 days you can set yourself goals that align to your end goal. By planning in detail like this you can then move with definite purpose. Each 90 days you can then do a very detailed plan with steps for each of the objectives for that quarter and times of when you are going to do them and how long each will take. This is an enormously powerful way to focus on your goals.

For example, after a 90 day review, your step-by-step plan for the next quarter could look like this (a blank one is provided in Appendix 3):

Quarterly Action Plan						
Step No.	Action	Time to complete	Date start *	Date finish	In diary **	Date comp -leted
1	List skills needed to get to Assistant Ecologist	45 mins				
2	Order Tree ID book and Flowering plant ID book	20 mins				
3	List 20 trees to learn – start notebook/scrap book for pressed leaves and notes	40 mins				
4	Get list of British plants, choose 3 families and start learning daily.	1 hour				
5	Book Phase 1 training course	30 mins				
6	Order Phase 1 handbook	20 mins				
7	Book protected species surveying course	30 mins				
8	List surveys needed for experience – species and numbers of surveys – prepare checklist to record progress.	1 hour				
9	Read Phase 1 handbook – put reading time into weekly diary	3 hours				
10.	Join local bat group	1 hour				

* Date start and Date finish is your way of deciding when you will do the action and how much time you will allow for it. If it is a quick job then put in the same date for both but if you don't do it on that day change the date or you may miss the step. Some jobs take a few days or even weeks. This way you can spread your actions out over the quarter and tackle them one at a time.

** Put the action in your diary so that when you get to that day you know what you have to do.
Record the date you completed it so you know you are on target and have completed all the steps for that quarter.

Look what you can achieve in around 9 hours spread over 3 months!

I use this process and every quarter I have up to 10 goals covering many aspects of my career and life and up to 10 steps for each one. Steps should only take an hour or two maximum, maybe only a minute or two, if they are longer than 2 hours cut them up into smaller steps. Each day I have planned to do something towards my end goal.

You can use this process to focus on all aspects of your life, for example: wealth and finance, health and fitness, family life, love life, fun and hobbies, personal development, and friends and social life. The 5% of high achievers in the world all have a written plan that they are working on!

Finally, although you have just started your first job your goals for your career might involve moving to another job if you are not able to gain the experience you need where you are. This can go into your 5-year plan. Don't forget though, there are also many ways you can get experience outside the workplace as we have already discussed elsewhere and in Chapter 5 in particular.

∞∞∞∞

Chapter 10

∞∞∞∞

Develop a Specialism

Although you will probably need to have a go at a bit of everything when you first start, eventually one wildlife subject will catch your interest and you will want to take it further. Many ecological consultants develop a specialism that is the focus of much of their career.

Developing a specialism might be a bit beyond the scope of this book, but, as we have just been thinking about goal setting, it makes sense to start thinking about specialising at the beginning of your career. This will help ensure that you will achieve everything you aim for. As long as you are aiming for it, planning it and consistently taking small steps towards it, you can eventually achieve anything you desire. You could prepare a Life Planner (see previous chapter and Appendix 3) for your chosen specialism and review it every quarter.

Some consultants specialise in a certain group of species, such as bats, and can make a comfortable living. However, to work in more diverse environments with a wider range of clients, and even to have a more interesting working life, I think it is good to have general expertise in many fields as well as an in-depth specialism in one thing in particular. In a team it works particularly well to have different specialist areas represented. I advise you avoid specialising too soon though – for instance I know a bat worker who knows no plants. She has always worked only on bats. This concerns me as she may not be able to recognise if she is in an ancient woodland (she cannot identify ancient woodland indicator plants), or even what species of trees are present, and this could be relevant for considering which species of bat may be present and describing the woodland itself for a report. For this reason I think getting a good general grounding to start with is essential. Become an ecologist with experience across species groups when you start out and specialise later.

Starting to specialise

If you have an idea of what you might want to specialise in, for example bats, you can start to plan your progress towards being a specialist straight away – but I recommend you do this without excluding growing your general experience in other subjects. Being a specialist is not something that you can achieve overnight. It might take a decade or more. You will probably need a couple of years experience behind you before you have a clear idea of what you want to specialise in. As you gain experience of more fauna and flora groups, and types of project, in your first job, you may be surprised by what you find most interesting. What might have been fun as a volunteer might be much less appealing when it's a professional responsibility. Likewise, you will get the opportunity to find out a lot more about our diverse wildlife and new groups will capture your interest.

Once you have decided where you want to specialise you can attend courses, work towards and gain your relevant licence (if appropriate), put in time in your own time, read up what you need to know, go to conferences to keep up to date, make contacts with other experts, and go out with them whenever possible. These things all take time and the sooner you start on the road the better. It took me nearly 3 years of bat work, courses and study to get my bat licence and at that point I certainly could not be considered an expert. Four years on from that I still do not consider myself to be an expert either, even though I have extensive experience (a classic case of the more I know the more I realise I don't know!).

Your path to specialising is up to you, you must take responsibility for your learning, put in your own effort and don't wait for others to direct your career for you or chase you up. I am now a trainer for bat licences and I heard someone complain recently that his trainer had not called him up to take him out on bat surveys and that is why he still hadn't got his licence! He had started working towards his licence at the same time that I did. Being pro-active and taking responsibility is the only way.

Who can help?

I found local mammal, bat, herp and badger groups were great places to meet local experts, get involved in local projects, attend events with specialists, listen to talks from experts and mix with like-minded people. Through these groups you will often hear of other events or courses available elsewhere or meet people who you can attend them with. Eventually it will be clear to you what you want to specialise in.

By volunteering for or attending local groups you can increase your knowledge and experience in your spare time. It is a great way to gain new skills and meet good contacts. Your ID skills can also increase under the tutelage of your fellow members.

Many of these groups either attend national conferences *en masse* or hold their own conferences. These are a great way to keep up to date on your chosen topic. Finding out about local initiatives through your local group can also expand your possibilities for getting experience and help you become an expert in your local area. As a beginner you also develop important relationships that can stand you in good stead if you are stuck at any time and need support or advice. I found this particularly useful as I set up my consultancy with no actual experience of consultancy. Fellow mammal group members where invaluable to me.

As I have said before, join your local societies, wildlife trust and other trusts so you can become informed of local initiatives, as well as having opportunities for training or other events and meeting like-minded people. You may find that you will meet the same people at several groups - the ecology world is a fairly small one.

Research

Once you decide on what to specialise in you might want to start getting involved in research. Helping others with their research at first to gain your expertise is a good way to start. Later, as you find out more, and realise what we don't know about certain species, you can start to contribute by doing your own research. Hopefully research can be done on the job, but maybe in your own time. Other members of your local groups may jump at the chance to help out and this gives you the opportunity to give back to the group.

If doing research appeals then I would encourage you to go for it. From your research you can write papers or articles in local and national magazines and newsletters, speak at conferences or local groups and pretty soon you will be a recognised expert. And then you will be the one that people ask to give talks, you will be the one that your colleagues will turn to for expert advice. You could even eventually be the one that the national organisations, government bodies or other national policy-makers ask to help them create policies and advice. You could influence or help shape how the whole country, or even the world, will deal with the subject of your specialism. Incidentally I know of national experts who are entirely self-taught and have never taken a relevant degree. With sufficient dedication and interest anything is possible.

Writing, talking, teaching

The logical progression from becoming a specialist is to write, talk or teach. Writing articles for magazines, newsletters, wildlife publications, scientific journals, the local press and of course books appeals to some consultants but not many, they already spend a lot of time writing reports. Once you start to write, however, you will automatically increase your reputation for expertise. There still remains an enormous opportunity for writers, particularly ones with an interesting or creative style who have a genuine knowledge or expertise. They can share their knowledge in a way that is interesting and understandable and they enthuse about the subject, which in turn gets readers interested in it. They spend the time putting together information that might otherwise remain un-written or at least un-sorted into a logical structure.

Of course once you are seen as the expert then you may be invited to speak to groups about your subject. It might be the local group of the wildlife trust or even a conference speech. There are many avenues for speakers but, as with the writing, it is unlikely that you will make much of a living as a speaker. I often get asked to speak at the local WI, luncheon club or to run an event for children. A young fellow I know from my university days, called Nick Baker, chose to go into the media after his degree and he has made quite a good name for himself on the television and writing books, however, there are not a massive amount of opportunities to do this, so to succeed you would need some extraordinary dedication and determination, which he has in abundance.

I know a couple of senior consultants who now do a bit of a speaker's circuit, including getting free trips on cruise ships in exchange for giving a series of lectures to the passengers. Some also run guided walks, wildlife holidays and whale watching on cross-channel ferries. Other people run classes, workshops or evening classes, but again it would be hard to make a living doing this but it does add interest to your working life. I have specialized in training people for ecological consultancy and offer many courses for the necessary skills and knowledge required to get onto the career ladder, other senior consultants offer specialist courses if they are a national expert and other consultants will attend their courses because of their reputation.

∞∞∞∞

Chapter 11

∞∞∞∞

Final thoughts

Ecology is an exciting and enjoyable career choice. You will meet lovely people, go to amazing places and see wonderful wildlife. What you get out of this career is up to you, what you sow you will reap in abundance later, when the time is right.

My final top tips are:

- Set goals and work steadily towards them. Keep them in sharp focus and ensure that they are time limited and specific. A goal should not be - *'In 5 years time I will be a senior ecologist'*, but rather *'By January 1st 2015 I will be a senior ecologist earning in excess of £40k, be an expert in all areas of consultancy and specialising in bats, I will be working with a happy, motivated team and enjoying every minute of my career'*. What a powerful vision and target that sets - much easier to focus on and work towards than the first example.

- Review your progress every quarter, plan the steps towards your end goal for the next quarter and steadily work towards achieving everything you have planned. Never give up - persistence is the key to success. If you are focussed on your goal, plan it and TAKE ACTION.

- Choose your attitude and make sure it is a good one. You will be remembered for your attitude – either as a good example, or a bad one. A positive attitude makes a massive difference to how much you enjoy your career and your life and how much other people enjoy being with and working with you.

- Take full responsibility for your life and career – no-one owes you anything, no-one has to help you. You need to create the life and career of your dreams yourself.

- Never stop increasing your ID skills. Being able to identify wildlife is the bedrock of professional ecology. You can always learn more.

- Aim high. If it is humanly possible then you can do it. National expert, your own consultancy, worldwide travel, whatever you can imagine in your wildest dreams you can achieve with planning, action and persistence.

∞∞∞∞

Good luck and let me know how you are getting on!

I would love to hear from you, especially if you have found this book useful.

∞∞∞∞

Sue Searle, 2010, Exeter, Devon.
suesearle@acornecology.co.uk

If you are interested in ecology courses visit:
www.acornecology.co.uk and
www.ecologytraining.co.uk

If you are interested in ecology courses and trips abroad visit:
www.pioneerecology.com

If you are interested in personal development visit:
www.thelifeyouchoose.net

Appendix 1

List of the laws we most commonly come across or need to know about for consultancy:

The Wildlife and Countryside Act 1981

Habitats Directive in 1992 and this was transposed into law in the UK as the Conservation (Natural Habitats, &c.) Regulations 1994 (The Habitats Regulations) – updated in 2010 – now Conservation of Habitats and Species Regulations 2010.

Countryside and Rights of Way Act 2000 – England and Wales and Nature Conservation (Scotland) Act 2004).

Badgers Act 1992

Natural Environment and Rural Communities Act 2006,

Hedgerow Regulations 1997

Town and Country Planning Act 1990 (as amended)

Town and Country Planning (Environmental Impact Assessment) Regulations 1999

Town and Country Planning (Trees) Regulations 1999,

Wild Mammals Protection Act 1996

The Planning and Compensation Act 1991

Environmental Protection Act 1990

National Parks and Access to the Countryside Act 1949

...to name but a few.

See **Naturenet** – lots of information about wildlife law see Countryside Law section
http://www.naturenet.net/index.php

International Agreements and European Directives

Ramsar Convention 1971 - The Convention on Wetlands

Bonn Convention in 1979 The Convention on the Conservation of Migratory Species of Wild Animals.

Bern Convention 1979 The Convention on the Conservation of European Wildlife and Natural Habitats

Birds Directive 1979 which set out objectives to protect birds that naturally occur in Europe.

Water Framework Directive 2000 concentrates on creating a better water environment with particular regard to ecology.

Environmental Impact Assessment Directive, introduced in 1985 and amended in 1997 looks at the impacts of projects on the environment.

Information on these can be found on JNCC's website:
http://www.jncc.gov.uk/page-1359

List of Protected species
European protected species:
- All bats
- Otter
- Dormouse
- Great crested newt
- Large blue butterfly
- Scottish wild cat
- Natterjack toad
- Sand Lizard
- Smooth snake
- Some aquatic species e.g. whales, dolphins, turtles
- Also some plant species

Other protected species that may be encountered on development sites:
- Badger
- All reptiles
- All *nesting* birds
- Water vole
- Pine marten
- Red squirrel
- White-clawed crayfish
- Some birds e.g. barn owls
- Some plant species

Protected habitats:
- Hedgerows
- Individual trees/ woods can be protected under Tree Preservation Orders
- Habitats that are used by protected species
- Habitats that fall within designated sites

See useful websites next to find out more about these laws and species.

List of useful websites:

This list includes a range of websites which you may find useful, whether for finding out about a particular species group or for information on wildlife law, latest news or for giving you information about particular survey sites. Have a read through the list and start browsing! It is amazing what is available online now!

Become familiar with the websites and what they offer. Knowing where to find reliable information is half the battle when confronted with a new challenge.

Note the web site addresses given here were correct at the time of publication but may be subject to change.

Wildife law

Naturenet – lots of information about wildlife and especially the law (Countryside Law section).
http://www.naturenet.net/index.php

JNCC website -
http://www.jncc.gov.uk/page-1359 - Conventions and legislation
http://www.jncc.gov.uk/page-1376 - UK legislation

Government organisations

Defra – Environmental and rural affairs. Good information about habitat management on farms.
http://www.defra.gov.uk/

Environment Agency – advice on preventing and treating invasive plants such as Japanese knotweed and giant hogweed. Excellent flood maps plus lots more.
http://www.environment-agency.gov.uk/

Joint Nature Conservation Committee (JNCC) – information on designated sites, Phase 1 Handbook download
http://www.jncc.gov.uk/ - home page

http://www.jncc.gov.uk/page-2468 - Phase 1 handbook

Natural England – find out about designated areas including SSSI citations, information on licensing for protected species, lots of useful info and a book shop.
Home: **http://www.naturalengland.org.uk**
Designated Sites:
http://www.naturalengland.org.uk/ourwork/conservatio n/designatedareas/default.aspx
Wildlife information:
http://www.naturalengland.gov.uk/advice/wildlifeguide/ default.aspx
Farming and wildlife information including habitat management:
http://www.naturalengland.org.uk/information_for/farm ers_and_land_managers/default.aspx#guidance
Licensing:
http://www.naturalengland.org.uk/ourwork/regulation/ wildlife/licences/applicationforms.aspx

UK Biodiversity Action Plan (UKBAP) – information on the national strategy for wildlife conservation including lists of priority species and habitats and information and action plans about individual species and habitats that are most at risk of declines or extinction. **http://www.ukbap.org.uk/**

Species specialist groups

Amateur Entomologists Society – Invertebrate Conservation and Study Society – lots of information on studying invertebrates plus keys to identification.
http://www.amentsoc.org

Amphibian and Reptile Conservation Trust (formerly HCT and Froglife) – charity helping to conserve amphibians and reptiles. Look at the Resources pages.
http://www.arc-trust.org

The Barn Owl Trust – local Devon charity with lots of good information sheets on all aspects of barn owl conservation.
http://www.barnowltrust.org.uk/

Bat Conservation Trust – home page – links to information, help and advice.
http://www.bats.org.uk/

British Dragonfly Society – ideas for projects and information about UK dragonflies.
http://www.dragonflysoc.org.uk/

British Trust for Ornithology – lots of information about birds including 'Birdfacts' about the different species, surveys to get involved with (including Garden Birdwatch) and the latest research into bird populations and trends.
http://www.bto.org/

Buglife – Invertebrate conservation charity – lots of good information.
http://www.buglife.org.uk

Butterfly Conservation – lots of information about British butterflies and moths including fact sheets on species, habitat management, projects and news.
http://www.butterfly-conservation.org

People's Trust for Endangered Species (PTES) – conservation charity with focus on endangered native mammal species. Including information about national surveys such as the Great Nut Hunt for dormice.
http://www.ptes.org

Royal Society for the Protection of Birds (RSPB) – brilliant website with lots of bird information including habitat management, gardens for birds and A-Z of British birds. http://www.rspb.org.uk/

The Mammal Society – information on projects to get involved with as well as excellent set of fact sheets on all

British Mammals.
http://www.mammal.org.uk

Other conservation organisations

British Trust for Conservation Volunteers – volunteering home and abroad including conservation holidays.

The Wildlife Trusts – information on wildlife, encouraging wildlife, events and links to local Wildlife Trusts. **http://www.wildlifetrusts.org/**

Habitat management

Farming and Wildlife Advisory Group – information about habitat management on farms for various topics **http://www.fwag.org.uk/**

Woodland Trust – links to several woodland management information booklets and leaflets. **http://www.woodland-trust.org.uk/**

Information about particular sites

MAGIC – shows designated sites and has datasets in the topics; Great Britain, Habitat Inventories, Rural Designation and Rural Land Base Schemes. **http://www.magic.gov.uk/website/magic/**

Nature on the Map - a Natural England website that shows designated sites and holds information about each designated site. **http://www.natureonthemap.org.uk/**

Old Maps - maps showing the previous use of the area. **http://www.old-maps.co.uk/**

Google Earth - aerial photography of a site to identify connectivity to other habitats and also very useful to see the buildings from the air. Street view can also be useful

for looking at buildings before you go to them.
http://earth.google.co.uk.

Other websites

Natural History Museum – information about British natural history and wildlife.
http://www.nhm.ac.uk/nature-online/index.html
Search for plant lists by postcode:
http://www.nhm.ac.uk/nature-online/life/plants-fungi/postcode-plants/

National Biodiversity Network – links to lots of databases for British species distributions and records.
http://www.searchnbn.net/

Appendix 2

Example of CV layout for ecological consultancy – see Chapter 7 – these sections should be included. How you lay them out is up to you. Templates are available on Word and Publisher amongst others.

Name

Contact Details – address, phone numbers, email address

Personal Information – nationality, DOB, driving licence

Summary of skills – bulleted list of skills – personal, surveys, personal skills, office and IT skills

Education – academic qualifications – most recent first

Employment History or Work Experience – list employment, without 'gaps'. State whether full or part time

Survey Experience – list any survey experience including any national surveys you have participated in.

Ecological Training – list any relevant courses you have been on (e.g. Acorn Ecology courses, local wildlife trust, BTO, FSC, NARRS)

Other Relevant Experience – list anything else relevant (e.g. membership of local groups, volunteering)

Achievements – list any awards or other achievements, especially if relevant

Hobbies and Interests – outline your interests so they can find out more about the real you. Don't make them up!

References – one from work and one from University or two from work. Work references should ideally be the senior member of staff or your line manager. Ideally they should have known you for at least a couple of years.

Appendix 3
Life Planner blank (enlarge on photocopier!)

Life Planner	
Career – Ecological Consultancy	
Timescale	**Actions/Aims**
5 years	
4 years	
3 years	
2 years	
1 year	
9 months	
6 months	
90 days	
Current	
Assumptions made	

Instructions: Write down your aim for where you want to be in 5 years time and write it in the first section under Actions/Aims. Write down your current situation and assumptions made. Then think of the milestones you would expect to be at for each of the timescales given and write them down. What can you do in the next 90 days to start your journey? You need to start NOW! Enjoy the process then TAKE ACTION!
Review every 90 days.

Bibiliography

See **www.acornecology.co.uk/shop** for a complete collection of suitable books. If we don't have the book you want let us know and we will add it to the bookshop!

Here are some suggestions to start your library:

Identification books

Flowering plants, mammals, insects, birds, trees, grasses, amphibians and reptiles – *Collins Field Guides* are excellent. Start with the subjects that interest you most – always useful to have a flowering plant one though.

Species ecology books

I love the *Whittet* books on various species for a great overview on ecology, field signs and conservation. Again, start with the species that interest you most or you are most likely to encounter in your area.

New Naturalist books also cover a wide range of species, habitats and related topics.

Also:

Badgers - Ernest Neal

Otters - Hans Kruuk

The Encyclopaedia of Mammals - David MacDonald

Mammals of the British Isles - Harris and Yalden

Available from: **www.acornecology.co.uk/shop**

Conservation Handbooks and Manuals

These are the 'bibles' of survey techniques, legal protection, habitat management etc. You will need to cite these in reports and have good knowledge of their contents. You can buy some of these, some are free to order as a paperback book but many are downloadable:

Dormouse Conservation Handbook – Natural England – order from NE: **http://naturalengland.etraderstores.com/Natural EnglandShop/IN29** or download **http://www.ptes.org/files/542_dormouse_ conservation_handbook.pdf**

Great Crested Newt Conservation Handbook – Froglife (free book) or download - **http://www.froglife.org/advice/gcnch.htm**

Water Vole Conservation Handbook - produced by the EA/EN/Wildlife Conservation Research Unit.

Lesser Horseshoe Bat Conservation Handbook – Vincent Wildlife Trust

Herpetofauna Worker's Manual – JNCC - **http://www.jncc.gov.uk/page-3325**

Bat Worker's Manual - **http://www.jncc.gov.uk/page-2861#download**

Handbook for Phase 1 Survey, JNCC, Peterborough. **http://www.jncc.gov.uk/page-2468**

Interesting reads - a must for your library

The History of the Countryside Rackham. O. (1986), J.M. Dent and Sons Ltd, London

Trees and Woodland in the British Landscape, The Complete History of Britain's Trees, Woods and Hedgerows. Rackham, O (1990) Phoenix Press, London.

Ecological Census Techniques, A Handbook, Sutherland, W.J. (1996) Cambridge University Press, Cambridge.

Managing Habitats for Conservation, Sutherland, W.J. and Hill, D.A. (eds) (1995) Cambridge University Press, Cambridge.

Enjoy building your library!

Don't forget to **read** the books
as well as collect them!